THE VELVET TOUCH

DEFUSING THE POPULATION BOMB

Kenneth Deboer

Copyright © *Kenneth Deboe*r 2024
All Rights Reserved

This book is subject to the condition that no part of this book is to be reproduced, transmitted in any form or means; electronic or mechanical, stored in a retrieval system, photocopied, recorded, scanned, or otherwise. Any of these actions require the proper written permission of the author.

Prologue

Many people are already painfully aware of the world's pressing environmental and ecological problems and at least a little depressed and daunted by them. Even more people, though, are either not sensibly aware of the 'realness', severity and immediacy of these existential threats, or they hope they will just go away if they ignore them hard enough.

So yes, we do bring very bad news. Thankfully, perhaps, we won't add any new previously unknown problems or new types of dangers to pile on to those already batted around. It is bad enough that our earth's overpopulation crisis is causing dangerous climate change and other ecological problems which are beginning to overwhelm the whole world's ecosystems as well as our own social wellness.

Our really bad news, though, is "it's worse than we thought." We, alas, actually are in real danger of partial or near total collapse of many ecosystems. We are well into the Sixth Extinction of life on earth, and it is by no means certain that the imminent ecosystem collapses that will occur will not eventually claim us as well. A recent UN report on "Biodiversity" calmly states that we will cause a million species or more to go extinct within a century or so. Unless—unless we humans do something!

Many books and articles recently have outlined what and how all this chaos and damage is being brought about. Many have also detailed how man, the God species, will soon have to take emergency measures to first slow, then stop their assault on the living world. At its base, the central problem boils down to too many human beings, doing too many things, using too much of our share of space, materials, energy, etc. We are indeed the cause of the wreckage in the natural world. This, increasingly, is preventing the whole biosphere from functioning as it has for the past several hundreds of thousands of years.

We will take the reader through the essence of the problem,

beginning with how Man has over-populated and come to basically lay claim to the whole earth and everything on and in it as his property. We will show what man's efforts to feed, clothe and entertain himself have done to the very earth he inhabits.

A litany of the horrendous environmental and ecological damage he has wrought will be outlined. Along with laying out the core problems leading to our present precarious status, we will also note some of the short-and long-term efforts which are needed to begin to solve them. Prime among these is the need to start to reduce the population of the world so that we can save not only so much of nature but us and our space as well. We will show why this is so necessary and how it could be done. We will outline some physical ways and processes that could largely make these existential threats soluble (or at least drastically ameliorated) by straightforward ways.

It would be noted that no new science, or even technology, or unknown mechanical, informational or physical types of equipment need be sought to solve this overarching problem. The problems are all human-caused, and almost all could be eventually cured by humans, basically with what we have now. Not at all easily, however. In fact, these are the greatest problems ever exposed in the history of mammals. It will require new ways of thinking, of governing, and even of thought. New economics, social structures and new understandings are needed; chief amongst these are societal governance and economics. Above all, we need to first call up the will to reduce the human population so that we not only can solve our current existential problems but prevent them from happening all over again. Again, the biggest problem is that of too many humans, which already far exceeds the carrying capacity of the earth.

An argument would be made that even if some reductions were to be made to our per capita energy, food, material and other needs, the ecosystem would still not be able to handle the load. We have gone well beyond sustainability already. We cannot solve our big problems bit by bit, or by humongous geoengineering, technical or scientific projects or even by simply a static population. We would then soon enough again be faced with the same problem as today. No government in the world has gotten a grip on any of these essential problems, nor are there any really serious, effective plans to solve them. Many groups and some countries are beginning to, and that is about the only encouraging activity we can point you to.

The problems we face, however, are all-natural, basic, and physical. Our main solution to the population problem, too, is simple:

reduce the population. As we will see, this could readily be done by known, standard means so that concomitantly, the huge, existential ecological problems will also begin to be solved or drastically reduced.

Reduced enough so that people could have full and expansive lives again that are also balanced with nature without destroying any other species' lives or habitat. Indeed, our call is nothing less than to make the most enormous, unprecedented leap to a change, a change larger than any ever made in our whole human history.

Contents

Prologue i

Chapter 1: Introduction 1

Chapter 2: Population Balance Sheet 8
    A. Population Trends Around the World 11

Chapter 3: Overall Effects of Man on the Biosphere: The Coming Catastrophic Decline 24
    A. Overview of the Natural World Damage 24
    B. Changes in the Oceans 31

Chapter 4: Biodiversity, Ecosystems & Humans 39
    A. More Bad News 40
    B. How and Why Are Ecosystems Being Degraded? 44
    C. Ecosystems Services 46
    D. So Where Do We Find Ourselves Now? 49

Chapter 5: Water and Air: Imminent Crises 52
    A. Pollution 70
    B. Greenhouse and Other Gases 72

Chapter 6: Land, Food and Agriculture 77
    A. Agriculture Footprint & Damages 79
    B. Big Monoculture Agriculture 82
    C. How Much Food Is There? 86

Chapter 7: Energy (Im-) Balance 93
    A. Overview of Energy 94
    B. Summary of U.S. Energy Use 96
    C. Energy in the Future 98
    D. Biomass 103

Chapter 8: Populations & Family Planning 110
    A. Short History of Populations & Family Planning 110
    B. What Should We Do About Growth? 117

Chapter 9: Population Reduction Plan 122
    A. Plea for a Cure 122
    B. How Much Should the Population Be Reduced? 128
    C. How to Do It 130

Chapter 10: To Build A Better World 140
    A. Where Are We, Again, & How Did We Get Here? 142
    B. Changes in Society & Government Needed 145
    C. More Suggestions for What to Do 149
    D. Last Words..152

Acknowledgments..158
About the Author..159
References..160

Chapter 1
Introduction

Dads sometimes advise their sons, "If you want to fix a problem, first find out what the basic problem really is, and then fix the cause of that." Well, we certainly have enormous, real problems. The main problem, however, is this: our recent global, increasingly intractable environmental imbalances are, for all practical purposes, derived in large part from and enormously energized by one underlying driving issue——overpopulation. That is the big problem that needs to be clearly recognized and fixed first and foremost. Both our evolved nature and our ever-increasing number conspire to wreak present and worsening havoc on Nature, if not also on human health and well-being——even our continued human existence.

In our daily lives, humans are born, grow up and raise our families in generally similar ways. We need clothes, food and shelter, social contacts, travel, and to entertain ourselves and each other. And we have designed, long ago, all sorts of physical edifices and artifices to satisfy these human desires (i.e. "Pursuit of Happiness"). For food and clothes, we've learned to farm and raise our domestic animals. For structures and materials that we like, we need to mine and harvest a plethora of nature's life and earth's bounty. We have needed to turn (some of) our water, air, and over-and-underground booty into new forms of matter, like plastic, fuel and fiber——and information! For social life, we feel an innate need to gather, and also gather up enormous quantities of wood and steel and ore and glass, etc., etc. (Man, it should never be lost sight of, is about the most social mammal there is; thus, we should not ignore such a basic biological fact, as humans have struggled these past few centuries to produce a better society and world.) This is what humans do. This is what we like. We like our friends and families and the planet and "things". A couple thousand years ago, our species invented civilizations. Now, we like the modern conveniences that these have invented, and, of course, see no reason why we wouldn't always have (some of) them. We pass our lives in and with these things and try to make our passage happy. That is all that any of us now living can do or have ever known. Why shouldn't we do these things?

We agree, we should. In fact, we can't help doing some of these things. It is in our deepest genetic nature, baked in through eons of evolution and millennia of cultural development. These are some of the worldly things that make our lives easier and, we assumed, ourselves happier. This is the "human game" of McKibben. And we have 'done good' with it. 'No worries.'

The problem is, the pace and scale of doing these civilized things have long since become unsustainable. Not simply unsustainable: dangerous. Dangerous for the peace and tranquility of our own lives, and dangerous for our natural surroundings and fellow inhabitants, and our own survival even. Simply by doing the normal human activities as engrained in our genes and passed through our cultures, which worked for so many centuries, we have generated the serious problems noted. We, our overabundant society of humans, are the cause of our major issues and the basis of those dire diseases that we are inflicting on our own nests. The disease has coalesced into one ultimate malady: too many of us doing the things we like and seemed born to do.

A few countries and groups have started addressing many of these in various ways; unfortunately, just desultorily and weakly, with no clear plans and definitely no clear majority will or direction. Hopefully, all these entities and more will keep on addressing these very difficult problems one by one in the future, too. However, most of us are not sanguine that this will actually happen in any meaningful way until better ways of articulating the problems and designing larger, more effective solutions are forwarded. A piecemeal approach to existential problems is like trying to cure diarrhea with a cork.

So the problem is us. The solution to the problem, therefore, is also us. The basic problem to be fixed, the core problem at the bottom of, or at least exacerbating so many of the individual, social and environmental emergencies besetting us, is simply that there are too many of us doing too many of these things. However, the earth simply cannot function the way it has for centuries, not with eight billion people on it. Not even five billion. Almost certainly not even two or three. The turmoil and energy and equipment and space requirements are too great.

Professor William Catton already, in 1982, when the population of the earth was about 4.5 billion, said: "We are already destroying many plants, animals and ecosystems with our Cornucopian myth of limitless resources."

Sir David Attenborough several years ago warned, (largely unheeded), that either we limit human population or the natural world will do it for us.

Here's the thing: we already grossly exceed the carrying capacity of the earth. It exceeds even the most minimal of definitions of human carrying capacity, i.e., the number of people that the earth could possibly support satisfactorily and sustainably. If that be so, then the logical solution to the problem would be to start to reduce the number of earth's humans so that eventually, each human will be able to do much of what humans do without drastic repercussions. Also, the earth's animals, plants and ecosystems could continue to be able to readily handle their vital jobs and function as they, too, were evolved to do. In addition, it should also leave us enough space and wild land for more satisfying lives for Nature as well as ourselves. And not overheat or produce other permanent harmful changes in the rest of the earth's biota and environment as we are so vigorously doing.

It should seem obvious, too, that trying simply to scale back a little or try a workaround for the damage being inflicted on our ecosystems, using high- gear production boosts, high-tech and all the rest, will not work. The energy costs and kerfuffle caused by us doing all these things will still exceed the humble ability of the biosphere to maintain stability under these conditions. (Not to mention that a lot of it boils down to simply an ever- expanding rat race.)

In short, we have already consumed, degraded, or destroyed too much of the biosphere. Enough of it already that our future, certainly one that is more, not less, satisfying, is not at all assured. The problem is gigantic; no larger problem has ever been faced since the appearance of ourselves on the stage. It is an existential one for us. We are simply using too much of our earth's limited resources to keep this game going for very much longer. Again, we have to shout-out the central fact that there are simply way too many people, who collectively over the past thousand years have led us to collapse of our livable environment. Most urgently, we need to come to understand that there are, indeed, 'Limits'.

Karen Shragg (2015) has for years been cogently urging us to "move upstream". That is, focus our efforts on fixing the cause of our first existential problem, not work exclusively on "downstream" solutions or remediations (like giant geoengineering projects, carbon sequestration, etc.). As Cafaro and Crist state in the Introduction to their excellent book Life on the Brink: "To ignore the population problem is to acquiesce in the continued ecological decline."

All of this news is not new. Many scientists, governmental officials, and ordinary people have been saying these things for several decades now. Our added prescription for fixing the basic problem for good is also not new or original. Our overarching prescription is simply this: start and continue to reduce the total population. The tools for the fix are already here and in place over much of the world. And let there be no doubt about the efficacy of family planning tools for curbing populations and thereby improving human living conditions. This has been clearly shown in many local demonstrations in various countries. The "trick" of population stabilization and/or reduction is astonishingly straightforward. Simply put, over generations, try to lower the birthrate slightly below the death rate. This is family planning, whether by education alone or by contraception or the panoply of other tools of the trade already available. We call it "The Velvet Touch"——and it works!

In the next few chapters, we will first attempt to accurately depict the central problem as it exists today and then argue for some simple, non-coercive means (almost all of which are currently in partial use) to help bring the earth into equilibrium again, as it was—— at least approximately——before the seventeenth century.

We hasten to add, that we are certainly not advocating that we simply turn the clock back or that society should simply revert back to the 1600s——or even much earlier! But just in our numbers. We want to build on our good deeds, like some parts of our civilized society, and retain 'enough' of our conveniences and standard of living. But, at the same time, we need to make it sustainable by not screwing up the functionings of Nature's business and our own well-being to boot.

Our first task in the next several chapters, then, is just to add a little updated information on particular aspects of the generally recognized problems and cite the long list of injuries being inflicted on the biosphere (which will eventually bite us, humans, too, some of these already have). Thousands of efforts, many of worldwide scope, are already in place, bent on solving each of the individual problems, like conservation, recycling, family planning and preservation parks for example. Without these efforts, the world would indeed be a poorer place. We need all these groups and efforts and more, much more. But without also a concerted effort to soon strike at the root cause of all of them, the task will be almost infinitely greater and ultimately hopeless.

We take it as a given (and hopefully everyone will sooner or later agree) that the world, i.e., man's activities, are making good progress in ruining or using up too much of the natural world. The systemic environmental problem of climate change is likely to be the first existential cataclysm to befall the human race. By itself, all (?) agree this is a bad problem. Certainly, there are many known ways to partially fix or partly alleviate some of the causes of this and some of the other world problems, too. For example, the general fix for global warming seems simple— lower carbon emissions. Some of the fixes for other man-caused ecological problems, like reef collapses, warming and acidification of the oceans etc., are beginning, thankfully, to be worked on, at least a bit.

Without, however, also reducing the scope and source of all these kinds of problems, the ultimate solution to these problems will be impossible. All of our severest problems, again, are related to, and mostly caused by, the real problem that lead us to this day—-too many people. Only fewer people doing 'people stuff' can truly solve our Nature problems, as well as some of our associated human problems. Then, for example, the climate change issues could become relatively simple to solve, completely and for good.

In addition to climate change (and in some cases, because of it), species are going extinct at dizzying speed from the additive effects of man's changes in every aspect of the world's biomes. That can't be good for the future of us or the biosphere; in fact, it will be fatal. Period! Last-ditch efforts are being made here and there to save the whales, the wolves, the tigers, elephants, rhinos, chimpanzees, sloths, and so many more. It will be sad to see so many of them go, but maybe there is time to save some of them, at least in preserves or captivity, initially.

Ecosystem after ecosystem is threatened in observable ways by other factors besides climate change, pollution, habitat destruction, etc. Scientists have detailed most of them in some detail. The root cause of all this is, again, too much digging and pawing in the biosphere at too great a scale. Adding even more people and technology to do more of the same is fortunately now seen dimly (or not at all) as not being a good idea and would actually make things hopeless. For at least the past century, population growth has continually erased the gains of much of our worldwide conservation and other environmental projects.

Some writers point out perhaps one hopeful fact: that the population of the world could be about to stabilize sometime later this century or next. That this process is already slowly beginning is a factor in our favor. Will it happen soon enough to even maintain any gains made in the various other programs for bandaging the biosphere? Our answer is: No. Not until strong, concerted social and governmental action is taken, and the population begins a significant decrease will the population explosion be over. Eventually, the number of people has to stabilize, willingly or no, at some much lower number than presently. It has to become understood that the number of people (i.e., the only species that has gained the power to totally remake the whole world) cannot exceed a safe and sane carrying capacity forever.

Some writers have even already declared victory over the whole problem. Bricker and Ibbitson (2019), for example, note the falling fertility and birth rates in many parts of the world. This, they say, will cause the world population to automatically stop growing. Soon, they say, the population will start falling and possibly keep falling (hence their "Empty Planet").

Most observers do not agree, and even if the population were to be stabilized at about the current level (which most experts agree is likely to happen within a century), that would not forestall the eventual apocalypse. Even then, we would still need to add more and more impossibly gargantuan efforts to greatly reduce the impact of our numbers and our works on the other acknowledged threats known to us. Continuing to try to solve all these problems, one by one, by the standard means, reminds the conservationist Tim Palmer (2012) of the classical cartoon. In it, a man is feverishly mopping up the floor with a towel from an overflowing bathtub—— but with the faucets still running full.

Admittedly, it is difficult for people to wrap their heads around our thesis here: that 'too many people' is, itself, our existential problem. However, that is the basic truth and is the first thing that we must acknowledge four square and get on with thinking about. This is the main reason for this book.

At the very least, population reduction itself would actually be the easiest as well as the key ingredient in combatting warming, soil degradation, habitat loss, extinctions and all of the rest of the things needing to be done to save the biosphere from further devastation. As we will try to show, population adjustment is actually quite easy and costs virtually nothing. It is a solution that, once started, can go on virtually effortlessly and unseen for as long as needed. This book, besides characterizing the threats to be surmounted, poses a guidepost for suggested population reduction and also some parallel practices to assure greater human well-being and to help soothe the way through the crises that await us. Just knowing that a potential satisfactory solution to the existential problem is underway should give hope and make people feel better.

And, again, a warning: it will do no good, as many people are now arguing (see, for example, Friedman, 2009), that since the population might be starting to soon stabilize, this alone would then allow us to simply apply all our standard, and new, fixes, and that would be enough do the trick.

"The alternatives are clear: A better life for fewer people or greater strife for more people."
—JK McKee

Chapter 2
Population Balance Sheet

In the next few chapters, we will briefly try to document the general conditions of our air, water, and plant and animal environment. These are the things that are increasingly causing fatal damage to the very fabric of nature as well as human society. We will summarize a little of the vast library of accumulated data on the status of our ecosystems to emphasize the seriousness of the problems we face with the deteriorating natural (and, thus, also human) environment. Most of these effects are due to overconsumption and many other of our common unsustainable practices. Ultimately, though, all of our dangerous 'overactivites' can be attributed to the sheer weight of our numbers.

Therefore, we will start with numbers. And, basically, only man's numbers count. The other biospheric members of the ecosystems usually stay in their lane. They haven't usually overrun their own niche or caused irreparable harm to their own species or more than a few local populations, at least for very long. (Large predators or invasive species, for example, may cause terrible and major harm to a few members of their prey species, but they do not usually drastically disturb all the rest of their ecosystem. Very few other species, except us, have been the sole cause of another's extinction.) It wasn't that long ago (like the 1800s) that tigers, bears, elephants, buffalo, lions and lots of other big, hairy scary creatures outnumbered humans by quite a margin.

Man's numbers didn't count much either, for his first few hundred thousand years. That is, while he was in his early years, even his early historical years. Even until several centuries ago, he went fairly lightly on the world stage. The world population 40 thousand years ago was probably only about a million or so *Homo sapiens*. By the time written history began, about 3000 BC, there were perhaps less than 25 million. The entire North American hemisphere had but 5 million people. But by Julius Caesar's day, undoubtedly the result of civilizations, the world population expanded over ten-fold, and there were already perhaps about 250 million. When Notre Dame was built, the population was around 320 million. In 1776, it was 775 million, and by 1804, the world passed the billion mark.

Thus, it took hundreds of thousands of years for the number of very smart upright apes to hit this one billion mark. But in the next 120 years, it had doubled to two billion (in 1927). As late as 1951, there were only 2.6 billion. And only then did the real explosion hit. In only thirty-three years, another billion had been added. Then, for the next five decades, about every fourteen years, we added another billion people. This almost tripled the population in just over half a century, leaving us with 8-plus billion today.

Incredibly, this happened under our noses, much of it in the lifetime of a great many of the people now living. We have been willing if unwitting, witnesses to the glories as well as some of the accidental gore accruing from this burgeoning population of extremely smart, energetic and adventuresome upright apes. Adding the next billion is expected to take just 20-30 years (a slight slowdown!) to make almost 9 billion by mid-century. Many estimates are that we will reach over 10 billion by the next century. Then, possibly, the population will begin to level off or perhaps decline slightly. The UN projection in 2022 estimated a peak population of 10,431,000,000 occurring around the turn of the century. Some other projections were that the world population curve might remain slightly upward until…? disaster, and who knows what?

Some people, like the futurist George Friedman (2009), for example, think that this would be a plenty good long-term arrangement, i.e., to wind up with a long-term stable population of 9 or 10 billion. He believes this would work to the advantage of the U.S., especially for the next centuries— provided we can get enough immigrants for our labor pool!

Looking now just at the U.S., by 1900, we had only 76 million people. In 1940, there were 132 million Americans. By 1951, there were 152 million total population, and by 2018, it had more than doubled again to 330 million. By 2100, barring massive catastrophes or changes in governmental policies, it could be twice that. (Significantly, however, all of our current population increase is due to immigration, so obviously, this problem will also have to be solved soon. Of course, if our population reduction measures were to be instituted worldwide, this immigration problem would become automatically resolved.)

It is useful to recall, which many of us who were living back then can do, what it was like, say, around 1940. Then, there were right about two billion people on Earth and only 132 million in the U.S. Those were the good old days, it seems. Our country seemed still largely rural (-ish), with much less jaw-dropping urban sprawl that so marked the next 80 years. It was not hard to find at least some open space between towns and cities. The environment, except in a few smoky cities, seemed pleasant enough, and apparently, most ecosystems (in North America at least) *seemed* OK. In a common perspective, we thought that life on the planet was in sync, and we needn't (as nobody did) worry about that. Growth and Progress were evident everywhere. Everybody 'knew' that this was just the natural order of things; of course, it was all common sense.

In hindsight, however, even in 1940, we were already well over the edge, and the damage that we were already inflicting on a gigantic scale was unsustainable and beginning to drastically show. We were just beginning, for example, to get over the shocks and disasters that unfolded with our self-inflicted poverty and agricultural mistakes in our interior with the Dust Bowl. We were (maybe?) dimly beginning to see that our human presence, energy, space and resource uses were already taxing or overtaxing the capability of 'things' to still function smoothly.

And a few scientists were already starting to sound the alarm. Among the earliest were ecologist William Vogt, 1948, (The Prophet in Charles Mann's book "The Wizard and the Prophet"), Paul Sears, and the conservationists WC Lowdermilk and D. Allen in the 20s, 30s and 40s and JK Smail, the Erhlichs and others in the 60s and 70s. Actually, voices were raised almost a century earlier than that. The pioneering Charles Marsh wrote the first book about the loss of animals and wildness ("Man and Nature") in 1867. Just a few years later, the first widely known advocate for the conservation of nature, George Bird Grinnell, wrote eloquently and forcefully, from 1872-1912, about the dangers of wanton 'unsettling' of the land. Richard Leakey and Roger Lewin (1977), almost 50 years ago, worried about the future of man and animals. Ehrlich's "The Population Bomb" in 1968 and The Club of Rome's "The Limits To Growth" in 1972 (Meadows et al.) were about the first hard smacks directly to the face.

Even well before the beginning of the twentieth century, our U.S. landscapes had already been greatly reconfigured, as it was bent to our needs and wants. Essentially, all of the Ohio and other eastern woodlands and wetlands were gone by 1850, for example. Agriculture had transformed most of the former wildlands into non-ecosystems. For another example, ten thousand years ago, there were an estimated 6.1 trillion trees in the world. In a recent estimate, there were only about one trillion. Humans are currently cutting down about 15 billion trees per year, so we invite you to do the math. Now, essentially, the whole aspect of the land is no longer functioning ecosystems; it is but Man's playground and work yard. It is, in fact, basically the finished product of hyperactive Boss, Man.

A. Population Trends Around the World

The population toll stands today at eight billion and counting. About seventy-five million more people are born each year than die. It would seem to be an elemental kind of arithmetical fact that if more people are born than die off, the population will grow, no matter what else is happening. This is the status of the population averaged around the world today. That is what it will generally look like, too, for the next decades unless something is done. As noted, there may be some little bits of good news, though. Generally, around the world, the population's growth rate is decreasing in a great many countries. However, the absolute number of people is still growing quite fast worldwide. It will keep growing until almost all countries have a Total Fertility Rate of 2.1 or less.

[Here, a few key definitions might be in order. Fertility Rate is the number of live births per woman. (A somewhat similar term, Birth Rate, is usually given as births per 1000 women.) Total Fertility Rate is a better measure of what's happening, as it denotes the total number of births per woman. Growth Rate is the number of births over the number of deaths. If that number is above zero, even 0.1%, say, the population will grow significantly. The replacement rate, i.e., the total fertility rate below which the population will decrease, is about 2.1 children per woman.]

For the past 150 years, all over the world, the net growth rate, births over deaths, had been well over 2 or 3 % or more. In the U.S., it was +1.3 % in 1980. Recently, however, the U.S. net growth rate has been falling and now is around +.1 %. Our total fertility rate is currently 1.7, i.e., slightly below our replacement rate. The positive growth rate, however, means that our population is still growing. (In our case, in the U.S., as in much of Europe and several other countries around the world, this is entirely due to vigorous immigration.) Worldwide, the average growth rate is 1.05 percent per year, and the fertility rate is 2.4. Both these numbers are down from all previous marks in the past half century or so, but still, these are leading to very fast worldwide population growth (seventy-five million per year).

Until 1970, very few, if any, countries had a negative growth rate. Now, a few dozen do, many of them small central European countries. Russia (population 144,500,000), Ukraine, Belarus, Bulgaria (population 7.02 million) and Latvia 'lead' with growth rates ranging from negative .05% to - .06 %. Germany, Poland, Italy, South Korea, Japan, China, Portugal, Hungary, and Greece also have either zero or slightly negative growth rates. Most other developed countries have, like the U.S., Britain, Denmark and so on, somewhat positive growth rates, but ones that could eventually lead to a stable population (if immigration could be ignored for now). Several other countries are already or soon to actually lose population, such as China, Italy, Japan, Thailand, Romania, S. Korea, Poland, Spain, Russia and a few others.

Actually, quite a large number of countries have fertility rates at or below replacement. If this continues, obviously, there will eventually (absent immigration) be a fall in population. In fact, the UN Population Fund recently reported that ninety-three countries have a TFR of 2.1 or below. The other ninety-five countries (or Territories) have a TFR of 2.2 or above, ranging as high as 6.8 in Nigeria. Obviously, these latter 95 countries are the ones that need to be the immediate focus of a population reduction plan (i.e., The Velvet Touch). In the U.S. and the other 93 countries that are on the road to non-growth, nothing really new needs to be done in the way of changes in "family planning". Their enormous new task for them lies in preparing programs for society and governments for the new age of non-growth that will be required to sustain it.

However, in the world as a whole, the human population is still increasing fast. This is mostly due to the large number of, especially Asian and sub-Saharan African countries, which continue to have high fertility and growth rates. Sudan, for example, has a 3.8% growth rate. There are several other sub-Saharan countries, such as Nigeria, Angola, Uganda, and Mali, over 3 percent. Ethiopia and Tanzania follow closely at 2.8 %, with dozens of other countries not too far below that. Most Asian and Latin American countries are not much better off. Campbell (2012) lists a few of the other countries with breathtaking (heartbreaking) population trajectories. Niger will grow from 16 million today to 58 million by 2050. Afghanistan will increase from 29 to 73 million, Pakistan from 185 to 335, and Ethiopia will grow from 85 million now to 173 million by 2050.

Ethiopia may be a good lab rat model. In 1950, Ethiopia had a population of 19 million. By 1984, there were 40 million (amidst a heartrending worldwide spectacle of a famine). Today, there are 85 million, headed for 174 million by 2060. Obviously, poverty, hunger, and misery have not improved in that country, nor, it seems to us, could it ever, on its present course. Nor is raising their standard of living up to near American standards a reasonable near-term option. As documented later, it would take several 'earths' to acquire the energy and materials to do that for all countries. Ghana is also often given as an example (of many) of an African, Asian or Latin American country with a similar dismal recent and current history. We will look later at the status of several other countries with equally bad populations and related problems.

So, instead of continuing to run faster just to stay in the same place as the plan currently seems, a middle option seems not only doable but intensely desirable. That is, lowering the temperature of the earth by reducing the source of most of the pollution and trouble, i.e., the number of people. As a key part of this, aid must be given to the various high TFR countries to gradually solve their existential problems with the numbers; this, in fact, screams out as a hair-on-fire emergency. Providing massive aid in every way possible, including assistance in planning, finances, agriculture etc., would appear to be an obvious and necessary goals. Relying on world trade to feed Nigeria's exploding population, for example, should begin to be perceived as a stupid idea.

Let's illustrate the numbers with some Central and South American countries. The table below shows a few of these country's recent population history and trends.

Table 2.1. *Population trends of a few selected South and Central American countries. (Numbers in millions.)*

Country	Pop.1950	Pop. 2010	Est. Pop 2050
Venezuela	4.3	27.2	40.3
Peru	7.9	29.9	38.6
Bolivia	3.8	9.9	16
Brazil	147.2	201.1	260.7
Costa Rica	.77	4.5	6.1
Honduras	1.2	8	12.9
Mexico	19.6	112.5	147.9
Guatemala	3.5	16.5	27.7
El Salvador	2.1	6.35	7

Poverty, hunger, displacement, overcrowding, crime and social and political unrest are the unnatural offspring of this kind of blinding and everlasting speed of growth in the numbers of people in all these places. Guatemala, for example, has a 4.15 Total Fertility Rate. It is no wonder that large numbers of people are desperate for any hope of escape from locally devastating conditions and unsafe neighborhoods.

Honduras, too, is another good example of how a burgeoning population is roaring apace, destroying their environment and their society. Kolandkiewiscz (2012) describes in agonizing and personal detail how their exploding population is driving wholesale destruction of not only their forest and other ecosystems but also a free-for-all in their whole government and roiling society.

To help these countries, then, and at the same time help alleviate our own problem with their fleeing, what is the best thing to do? Build a Wall? Send in food, goods, money, medical and structural aid? The Army?

Some of these would be fine. But along with all other aid to fix their basic problem, we might suggest sending in an army from Planned Parenthood! Stabilizing and lowering the population of each exploding country is the only practical long-term solution. For this, all that is really needed is standard family planning (The Velvet Touch). Organizations like Planned Parenthood and dozens of others have all the expertise, materials, wherewithal, knowledge, capable personnel, and everything else needed to reduce the number of children born each year in Guatemala or anywhere else.

We don't mean, of course, to literally send Planned Parenthood (or the army!) down to each country. We mean, countries should use the various governments and other relevant agencies in their countries to consciously lower and eventually stop population growth. It is abundantly clear from data from several countries around the world that using the well-known educational, training, technical, and advising methods and assisting with the modern methods of contraception is all that is required to slow and stop runaway population growth. Think of it as the centerpiece of a "Latin America (and World) Marshall Plan."

We should be happy to assist all governments in this effort. Of course, the results, the payoff, would not begin to be seen until a decade or two in the future. But letting some of the air out of the existing situations should prove of some immediate value, just knowing that things will get better. This is really a very simple, inexpensive, straightforward way to start making a difference for people and programs right away in a large number of countries (and also to eliminate the immigration 'problem', too, worldwide). These are examples of the kind of prescription we are advocating for the whole world. This is the basic heart of our population reduction plan: i.e., the Velvet Touch. The basic problem, of course, is not "A Border Problem" as so many loud politicians claim; the issue lies in the ever-increasing population explosions and the resulting chaos in the aforementioned countries. Their problem cannot be solved by some magic policing or border wall construction. The real challenge is the burgeoning populations, overcrowding, and decay in these countries. The only sensible way forward is to assist these nations in addressing and resolving the root cause of their problems, rather than seeking retribution against the desperate migrants who are fleeing their misery.

So, is the earth sustainable now? Not a chance! The population today is definitely not sustainable. In 1798, Malthus pointed out that species' populations tend to increase exponentially, but that environmental largess, including providing enough food and resources, is finite and relatively fixed. That brought about the specter that has haunted our dreams ever since—— people dying by the millions and billions of hunger.

But the Malthusian doctrine that has been mocked and/or feared for so long turns out to be not so simple. It is true, as 'realists' have touted for so long, that human populations will not necessarily be unable to feed themselves, possibly even at the levels needed by 2050. The tremendous efforts by governments, science, agriculture, and cultures to increase their food supply have indeed been quite successful. As detailed in later sections, we actually could (maybe) find ways to produce enough food to feed the world for a short time to come. The problem, however, is that the cost of doing so is too high for our natural world to maintain itself——and us.

Let us now, then, summarize the current status of populations overall.

Figure 2.1. *below is the most important graph of all human history. It shows the exponential growth curve of the number of humans from 6,000 years ago until today (i.e., all of human written history.)*

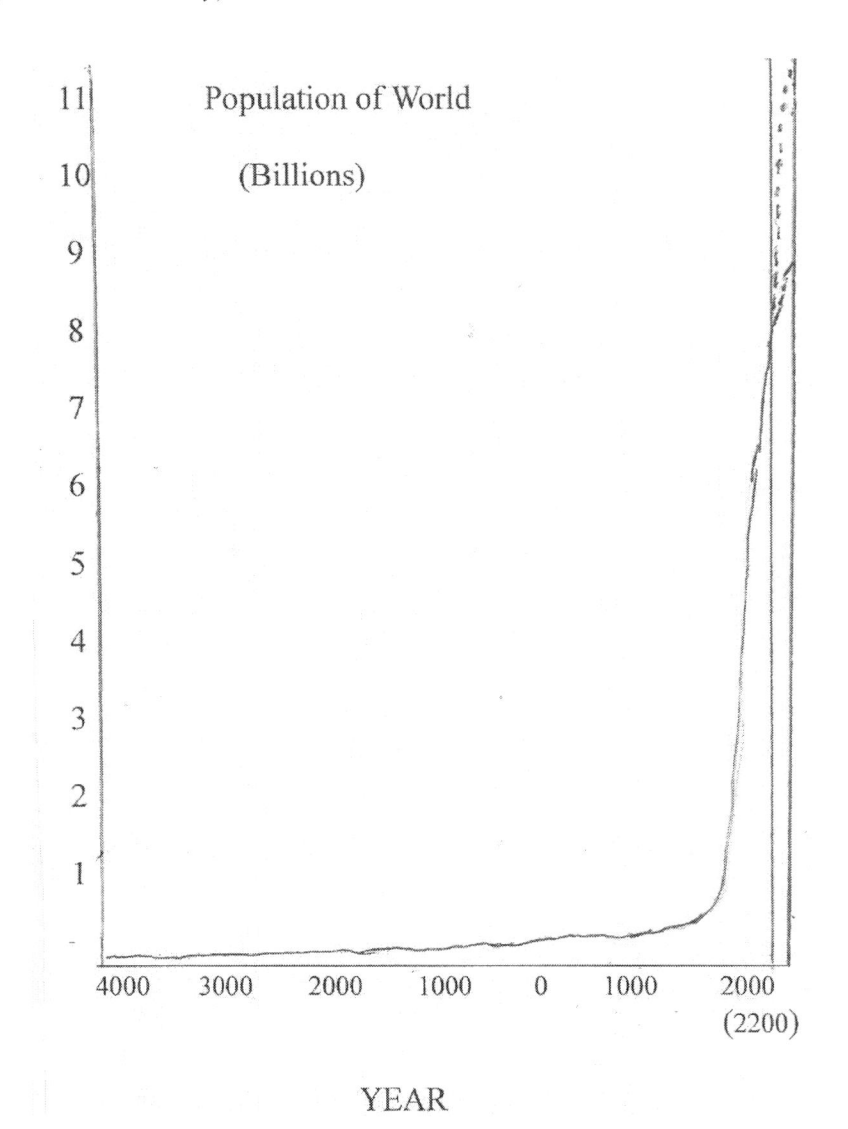

Population of World (Billions)

YEAR

As Figure 2.1 shows, there were essentially no people for millennia until the beginning of history, six or seven thousand years ago. For hundreds of millennia, the population grew very, very slowly, numbering only in the thousands worldwide, even up until 60,000 years ago. Even by the time of Caesar at the beginning of the Present Era, there were only an estimated 200-250 million. Until then, the ecological footprint of the human species was quite light in most places.

By Caesar's time, and almost a thousand years after, the number of people rose still very slowly but steadily until about 1700. Then the curve takes a clear upward trend, steadily, until the Industrial Revolution. Then the explosion hit, first with a marked spike, then this spike shot straight up to the sky. Doubling, then tripling in less than a century. No other large species growth ever did, or probably ever could, produce such a sustained straight-up curve. We are witnessing now the crushing, ever-increasing effects of that explosion.

The estimates of what will occur for the rest of this century and the next are also scary. What some experts have estimated could be the trajectory of the population in the future is shown by the extrapolated dotted lines. These dashed lines represent High and Low population estimates.

As the extrapolated (dashed) lines in the Figure show, most of the estimates are that the world population will keep climbing, albeit more slowly, until around 2100. (Or until a catastrophe somewhere in the biosphere sends it and the human population into a downward spiral.) It should be noted again that there is an expected flattening, especially in the 'Low estimate' curve, sometime before 2100, when the earth's population would stabilize and/or decrease a little. Most current writers would agree, though, that even if this dip in growth does occur, the galloping ecological disasters will continue on anyway, maybe just not quite as fast. We've already gone well beyond any safe population.

As noted, practically no matter what we do, barring aggressive action of the kind we are advocating, the earth will have at least 10 billion people, and probably more, by the turn of the century. The most likely trajectory after that is that the population will first grow more gradually, then possibly followed by stabilization. Then, presumably, numbers would slowly decrease as culture and education develop (or, more likely, disaster strikes). Some estimates, however, have plotted little flattening of the curve at all and no stabilized population for the next two centuries (Rakesh, 2014). In either case, almost all ecologists agree that ecological (and human) catastrophe will be the final result.

Next, in Fig. 2.2, we will look at the situation in the U.S. alone and for comparison to the world's population growth.

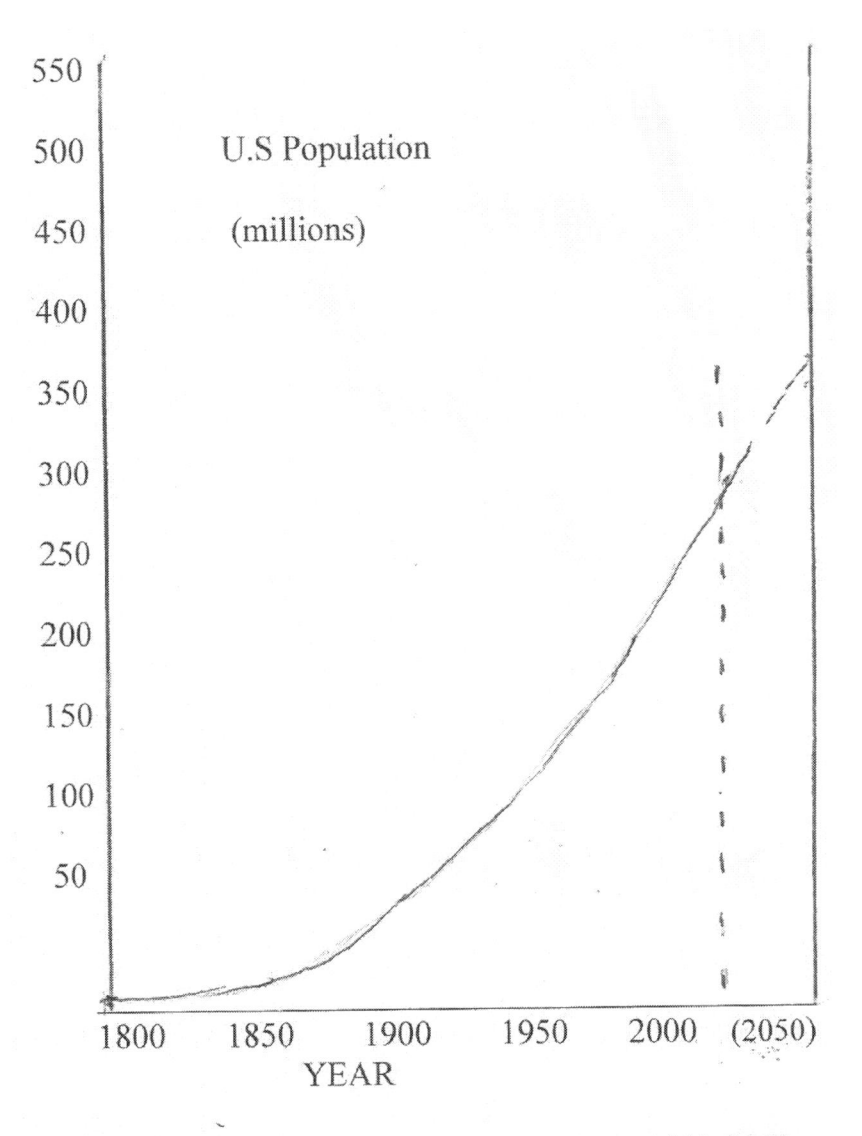

Figure 2.2. Number of People in U.S. from 1800-2050.

Figure 2 shows the U.S. population (in millions) since 1776. The first thing you notice is that the population curve for the U.S. looks very similar to the world (Figure 2.1). The growth rate starts slowly. By 1776, there were only 2.5 million. Despite continued immigration from European countries and elsewhere, there was a quite slow increase even to 1830, when there were only 13 million. Then, the explosion began. Slowly at first, to 1900 when there were 76 million, then faster and faster, accelerating all the time, right up to today. The population explosion was beginning to really bloom by the year 1927, when it hit 119 million. Exponentially, the population rose again to 152 million in 1950, 216 million by 1974, and 288 million by 2000, continuing the steep curve up to today.

As noted earlier, the U.S. birth and growth <u>rate</u> is now slowing hugely. However, the population <u>number</u> is expected to continue to still go up, albeit less rapidly, until at least 2040 and then (maybe) begin to slowly level off. The number of births over the number of deaths last year in the U.S. was the lowest ever in our history, about 0.1%. In 1960, the population was 171 million, but the TFR was 3.6. This produced 4.3 million births and only 1.7 million deaths. There were also 330,000 new immigrants. By 1990, the population was up to 250 million, but the TFR was only 2.08. Nevertheless, there were 4.1 million births to 215 million deaths (plus 1.3 million new immigrants). Now, (in 2020) there were 3.6 million births to 3.3 million deaths (plus 1.5 million new immigrants. Thus, it is owing to immigration that the population is still increasing. Fixing this is one of the world's great initial tasks. [On September 7, 2023, at a press conference in New York City, the Mayor stated that if we don't soon do something about the immigration problem, "Immigrants will destroy NYC."]

The key part of this is to start changing mindsets so that lower populations elsewhere throughout the world come to be seen as a necessary goal and that growth itself has to be permanently banished.

Many experts project that there will be a further slowing, stabilizing and possibly a slow decline before the end of this century. Some experts doubt that it will actually go into decline but would simply more slowly keep increasing, at the least to 440 million and possibly 600+ million or more by 2100 (Ortmann & Guarneri, 2009). In Fig. 2.2, the dashed line beyond 2020 illustrates these estimates. Several estimators see a stabilization of population (depending on the dicey immigration situation in the future). Until disaster strikes, however, very few people see the population falling to any significant degree. The very best outcome anyone actually projects is eventually simply a stabilized population at about our current level (which will not help us out of our existential jam).

Ok, these are the basal numerical facts about the size of Man's increasingly massive presence in the world yesterday and today. Up until a few millennia ago, they were more or less irrelevant to the biosphere at large. Local populations, to be sure, did cause some considerable amount of trauma to the local fauna and flora. In fact, they caused or had a hand in the extinction and near extinction of several of the most vulnerable of local animals, for example, the miniature rhino, dodo bird, passenger pigeon, moas, Tasmanian tiger, short-faced brown bear, saber-toothed cat, Great Auk, and others (Dirzo et al., 2014). From 1600 on, man has been claimed to be responsible for the complete extinction of 680 species of mammals alone.

Today, man's numbers are everything. Impacts on every aspect of Everyman's daily existence and his total needs are amplified to the nth degree. The numbers reverberate around the world, affecting every person and every major plant and animal group. Importantly, they impact the instantaneous operation of every facet of our physical surroundings: our air, water, soil and boundaries. The water cycle, the nitrogen cycle, the carbon cycle, weather cycles, ecosystem balance, and life or death of frightening numbers of our earth's animal and plant neighbors are all massively impacted all the time by the number of men and their roiling, ever-growing activities.

In the past 50 years, while the population has more than doubled, the global economy has more than doubled, and world trade has increased tenfold, driving up demand for energy and materials. Thus, the individual footprint grows, too. Even worse for the environment, the footprint actually is growing faster than his raw numbers. Thus, does Homo giganticus now roam the earth, foraging. In the future, as the population grows, or would even stay the same, the problem grows—until it doesn't. And then everything starts to fall, in a chilling scenario that now, apparently, only scientists see clearly, and movie makers and sci-fi authors dramatize. There are just way too many people, collectively doing too much damage to the environment—i.e. our own two-thousand-year-old wrecking crew. Thus, our created unsustainable conditions in the whole biosphere already is an imminent existential problem.

We conclude this chapter with one more chilling chart to chew on. (Fig. 2.3)

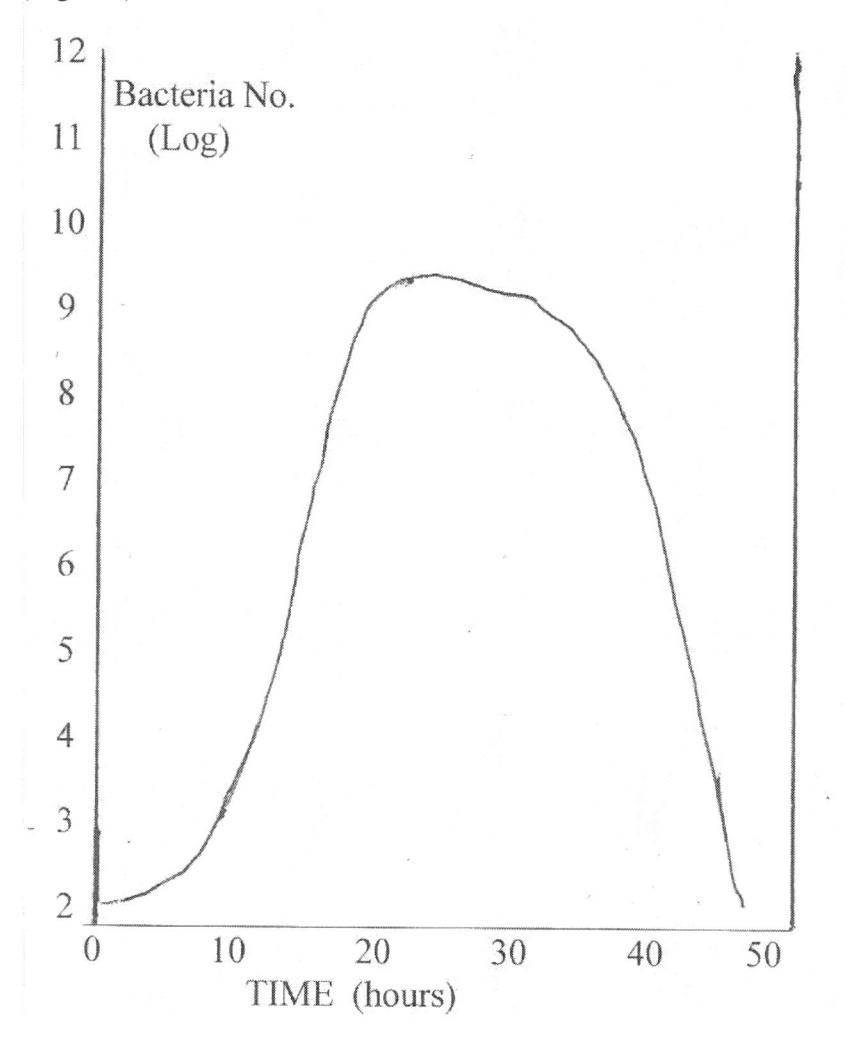

[Graph: Bacteria No. (Log) vs TIME (hours)]

Figure 2.3 shows the growth curve of populations of bacteria (virtually any bacteria) that have been placed onto a petri dish. The dish contains a finite amount of all the nutrients that bacteria need to grow and thrive. At first, the bacteria love it and chomp down and do what bacteria (and most other organisms, too) do best—reproduce. Every hour or less, there are ten times as many bacteria happily chomping away. This goes on for several hours as the curve goes up exponentially.

Then, suddenly, the good times end as the crowding grows. The food is just about gone, and some of the metabolic chemicals begin to become a bit toxic. Reproduction almost stops, and dying begins. Slowly at first, then reproduction stops and dying continues big time. Straight down, to oblivion. This is life in a finite, contained space in a nutshell. There is no way out of obliteration in these, our present situations.

You can hem and haw, and haggle and shuck and jive all you want, but in the end, the largest, central problem facing us will be seen to be the population of man. That is the driving central problem that underlies and exacerbates most of the nagging, and sometimes dire, situations facing us, with oceanic degradation, pollution, global warming, not to mention all the ancillary wars, poverty, refugees, governmental failures, etc. etc. Overpopulation lies as the proximal cause of our biggest ails, i.e., the harms to our natural and social environment that has been and is, more and more rapidly, happening. Overpopulation, therefore, is the main problem we will address ourselves to find solutions to. It would seem that if we began to solve that problem, then the unprecedented problems produced and/or exacerbated by this ultimate cause should increasingly become much more tractable.

"The human growth curve is more bacterial than primate."
—EO Wilson

Chapter 3
Overall Effects of Man on the Biosphere: The Coming Catastrophic Decline

A. Overview of the Natural World Damage

We can start our tally of Man's impact on the biosphere with general considerations about man's custody of the land and sea. The first thing everybody learns in Ecology class is that "everything affects everything else." Essentially, each species of plant and animal is in constant interaction with its environment and each ecosystem of biota interacts, eventually, with everything else, i.e., every other ecosystem. The biosphere is One. The earth is One. Therefore, if several parts of an ecosystem become lost or degraded, the whole will eventually feel exacting effects. Extinction of even a single species, then, could have powerful effects on the whole system. (Unfortunately, most people either don't know this or just plain don't believe it——or maybe don't care or have time or inclination to think about it.)

From even a cursory look, it is clear that the impact that Man is having on the earth is already causing us to enter the 6th extinction. The onset of this mass extinction, far above the rate of most previous extinction events, was noted over forty years ago by scientists like Niles Eldredge, JK Smail, Paul Ehrlich, Leakey & Lewin, 1996, and others. Elizabeth Kolbert, in her recent book, "The Sixth Extinction," points out some of the negative impacts that man has already made in generating a rapid extinction rate. She describes some of the best-known examples of species extinction and how this is happening. She states: "Right now, we are in the midst of the 6th extinction, this time caused solely by humanity's transformation of the ecological landscape."

Richard Leakey's 1996 book, "The Sixth Extinction," as well as Terry Glavin's 2007 book, also called "The Sixth Extinction," were also early 'cold water to the face' alerts.

Mass extinctions are the last thing one would think could happen because of any one single rogue species. In the history of the earth over the last half billion years, it has rarely, if ever, happened. Until now, there have been only five mass extinctions, the first one 400 million years ago and the latest only 65 million years ago. That one, at the end of the Cretaceous period, essentially happened in one day and its long-lasting aftermath. Over 50% of all land life became extinct in a relatively very short time. A huge meteorite crashed into the peninsula off the present-day Yucatan peninsula of Mexico, and the fires and dust pall spread all over the world, lasting centuries and causing more extinctions the whole time. The most famous casualties were the dinosaurs, of course. We are one of the direct descendants of the first little mammals that had survived that mayhem.

However, now we are again into another big extinction event, as species after species are dying out just about as fast as after the meteorite disaster. But this time, our species is causing almost all of it. Thousands of documented species have been extinguished in the past 15,000 years, partly at the hands of man, but the current rate is hundreds of times faster than normal. And the pace is only quickening. Almost all of Earth's large mammals are at grave risk of going extinct, and many of them are already extinct in the wild.

In North America, our early Asian human transplants had a hand in carrying out, in a fairly short time, a number of extinctions. The Mastodon and wooly mammoth vanished around 10,000 years ago. The saber-toothed tiger, short-faced bear and others were also early victims of the newly arrived Americans. The Great Auk, by 1844, was one of the last ones and was definitely a straight-out mass killing. So was the passenger pigeon. Even a bit before the sixth extinction was recognized (around 1980), a goodly number of the earth's large mammals and 15% of fish had vanished since the dawn of man some 200,000 years or more ago. Probably, the last little, wild American buffalo herd was exterminated by Indians, ranchers and buffalo hunters near the Musselshell River by the Snowy Mountains in central Montana in 1886 (Punke, 2007).

Most of the large or iconic species are now under great danger of extinction. The Sumatran rhino, Northern white rhino, black rhino,

Asian elephant, Malayan tiger, orangutans, gorillas, and several species of whales and lemurs are some examples of iconic kinds of animals very threatened or endangered by extinction (Kolbert, 2014; Crutzen, 2002). The iconic koala bear and panda are also getting close to extinction in the wild, as are, in fact, all big mammals. The Caspian tiger and Javan rhino are believed to be already gone. Gone also are Xerces and dozens of other butterflies. The majestic monarch butterfly is over 90% reduced and in grave danger. Dozens of other species, like the prairie chicken, wolf, black-footed ferret, many plants, and so many others, are also virtually extinct, mostly kept going in parks, preserves, or captivity. The ancient and beloved Joshua tree in the American West is rapidly declining due to climate change and other encroachments. It is predicted to be extinct in the wild by 2070.

The Nobel Prize-winning Paul Crutzen cites especially chilling work with the Sumatran rhino. It is extinct in the wild and will probably go completely extinct, despite heroic efforts to keep alive the dozen or so individuals still kept in captivity. Hunting and poaching are virtually the sole cause of its demise. This is similar to the fate soon awaiting virtually all of the existing megafauna, especially all species of rhinos, elephants, hypos, the big cats, most apes and similar.

Besides the immediate threat of extinction to thousands of species, the rapid decrease in the numbers of <u>individuals</u> in thousands of different species in many disparate ecosystems is doubly alarming. Already, we have caused a huge decrease in the number of individuals within species. Overall, in just the last half-century, wild marine mammal numbers have fallen 8 percent. Bird numbers are down over 30%, and all mammals as a group are down 25%. The Audubon Society recently reported that 389 of the original 700 or so native U.S. bird species are in danger of extinction. Perhaps the hardest hit group, though, has been amphibians, down over 40%. Turtles and lizards are down almost that much, and there has also been about a 40% decline in insect numbers worldwide (think twice before saying "Good"!) Not only animals but of the hundreds of thousands of known land plant species, the numbers of a great many are down nearly by one-half.

Even a brief listing of the damaged ecosystems and the gone or threatened species is frightening.

- Less than 50% of the original tropical forest is left.

- East Africa forests cut down 90%, West Africa >75%.

- The U.S. lost over 90% of old-growth forests.

- 85% of wetlands around the world have been lost

- Human activity has highly transformed at least 75% of the land surface.

- Most of the world's major rivers dammed or diverted; many run dry much of the time

- Fisheries are now removing 1/3 of primary production of the oceans.

- Humans use more than ½ of the world's readily accessible freshwater supply.

- 57 species of fish went extinct in the U.S. in the last century.

- 40% of the earth's extant plant species are endangered due to pollution, pollinator loss, land use changes, climate change, and deforestation. Desertification, livestock, traffic, etc.

- In the Mobile River Basin alone, 37 species of aquatic snails became extinct due to dams, reclamation, irrigation, etc.

- 19 species of mussels have become extinct in the U.S. over the last 3 decades due to impoundments, silting, pollution, etc.

- Numbers of endangered mammals 75%, reptiles 65%.

- Endangered amphibian species, 55%, birds 45%.

- Coral reefs, 19% are already dead, and 38% of known coral species are direly threatened by extinction.

- 32% of amphibians are forecast to be extinct by 2050.

- The African elephants, if present conditions continue, will be extinct in the wild in 30 years.

- Number of studied crustaceans is down 27%.

- U.S. conifers numbers declined 34% within the last 50 years

- 41% of insect species are down by ten percent or more in the Puerto Rican rainforest.

- Flying insect numbers are down about one-third worldwide.

- 84% reduction of the burrowing mayflies in Lake Erie in the last decade.

- 77% more fires (80,000) burned in the Amazon rainforest in 2019 (mostly due to agriculture) than in any previous year in history.

- East Antarctica glaciers have passed a tipping point.

- Summer sea ice in the Arctic is predicted to disappear by 2050, endangering the remaining 900 polar bears.

- Bees, moths, hummingbirds, bats etc. (i.e., the pollinators) are declining rapidly.

*Pollinator losses, due to the decline in bees, butterflies, bats and other insects as well as birds, are an existential threat all by itself. As mentioned, the monarch butterfly in North America is down over 90% and in great danger of extinction due to habitat loss of one group of plants, the milkweed. A major cause of most of these losses is the widescale use of pesticides and herbicides for the past three-quarters of a century, which is catching up to us now, as Rachel Carson warned 60 years ago. Pollinators play a crucial role in reproduction in 80% of the main food and fiber crops around the world. There are at least 7000 nut and berry crops, for example, that depend on these species. The main pollinators are about 100,000 species of insects alone, chiefly bees, wasps, moths, beetles, etc. There are also about 1500 species of birds and mammals that are an essential part of the crew as well. Bats are one of the important mammals in this group, and most of the common perching birds play a big role in ecosystem pollination. A recent Cornell University study estimated that crop pollinators contribute 29 billion dollars to U.S. farm income annually.

The International Union for Conservation of Nature (IUCN) keeps the most up-to-date census of endangered or threatened species. Their current tally of 41,000 plants alone (and 29% of the known ones in the U.S.) are now on this roque's list. That list rises quite fast every year; in 2019, it rose to 105,000 species. The 2019 UN Report on Biodiversity predicts that a million or more species will become extinct by the next century if current trends and practices continue (which seems very likely). What is now happening has never happened in the last 65 million years, before the Age of Mammals. Over half a million species on land have insufficient habitats for long-term survival and are likely to go extinct within decades unless habitats are restored. One in four species, plant and animal, are endangered already. The last major extinction, 65 million years ago, occurred long before most mammals or, indeed, very many of the currently familiar plants and animals were around. This time, we are losing the cream of the crop——the last and highest evolved forms yet.

The total land area of the world is about 57,306,000 square miles (36.8 billion acres). A little less than half of these, or 15.8 billion acres, are classed as habitable. A Geologic Society of America Report, 2018 found that:

"People have directly transformed more than half of the area, 27 million square miles (17 billion acres). Most of this development is by conversion to cropland and pasture, etc., plus cities, shopping malls, logging, mining, roads, sprawl, and all the rest of human actions on the earth. Various authors have calculated that humans live on only 10% of the land but occupy or use, in some significant ways, almost 90% of habitable land."

On the 50 million-plus square miles on earth that are ice-free, there are 8 billion people. That means that we already have 160 people, on average, in each square mile. That number alone would indicate strongly that Man is squeezing out the habitat for much of the world's inhabitants, plant, animal and microbial. Unfortunately, the lack of suitable habitat is the most significant factor in the drastic decline in the numbers of plants and animals, as well as the outright extinction of so many whole species.

It was recently estimated that the coming sea level rise alone will cut the range of all earth's plants by 18 percent. That will also, obviously, drastically decrease the range for the large vertebrates, too, by 8 percent. (That means 'You too, homo, the wise one!')

Foreman, Crist, Cafaro, Wilson and many other conservationists and scientists have given excellent and extensive accounts of some of the damage that is befalling virtually every kind of ecosystem worldwide.

Foreman (2007 calls our present way of life of scraping away wildness "landscaping".

Prof. Eileen Crist (2012) says, "Overpopulation has been driving the dismantling of complex ecosystems and native life and leaving in its widening wake of constricted environments, simplified ecologies and lost life forms…Non-human genocide by means of which the Earth is 'put to work' 24/7 to serve a master race… Nothing seems to sway the global social collective from its presumptive intent to constitute the entire planet as a human resource domain."

From the works of these authors and many other sources, including the 2019 UN Report of Biodiversity, the following is another short compilation of some of the evils and problems noted occurring around the world:

- Overpopulation is the cause, and also the result, of upheaval, social, economic, political and industrial resources.

- What kind of world do we want to live in? Natural limits are important to incorporate into our conscious brain.

- 2 out of every 5 acres of land is now in crops; we need to redefine carrying capacity to include the natural world

- 2.2 million non-urban acres per year in the U.S. cleared, paved, subdivided, built on, bulldozed

- Since 1700, the percentage of land going to feed man went from 10% to 40% or more.

- 70+% of freshwater goes for irrigation.

- The sea level is likely to rise 9 meters by 2100.

- Crops and pastures cut up and fragment wildness.

- Urban development and cities and similar land appropriation take up nearly 3 million more acres in the U.S. yearly.

- Agriculture, including forestry, has been and is the main source of habitat destruction; plus, spraying adds to the spreading of damage; one-third of the world's cropland is seriously degraded.

- Desertification, especially in N. Africa, is a major and increasingly devastating problem.

- Energy use and habitat destruction are related forces of impact.

- Most activities that lead to indirect endangerment are energy intensive, i.e. in 1981, humans coopted 25% of all photosynthesis, now it is nearly 50%.

- We are now radically changing the global Carbon and hydrological cycles.

- Most of the native plant cover currently left around the world is in rainforests and agriculture monocultures.

- Summer sea ice in the Arctic will be gone by 2035.

B. Changes in the Oceans

Besides the great adverse effects on the land, the damage is just as bad in the oceans (Mitchell, 2009). The oceans and other large bodies of water (the so-called aquasphere or hydrosphere) comprise about 80% of the 510 trillion square meters of the Earth's surface.

Notably, this is also where nobody lives, so one might first think that it would not suffer much from the effects of Man on land. But au contraire. The oceans may turn out to be the first to be catastrophically affected. The Earth's oceans are actually the main regulators to maintain stable temps as well as our atmosphere worldwide. Oceans act as the main energy storage reservoir and controller over the whole earth system. David Trexler categorically states, "What happens to the ocean happens to the Earth in general."

The North Atlantic Ocean temperatures have risen to record highs by 1-2 degrees C the past several years——and this keeps bending rapidly steeper.

Air and marine pollution have increased tenfold since 1980. The core of the ocean's ecosystems, the phyto- and zooplankton, are being hammered by acidity, pollution and rising temps and many of the ecosystems are rapidly decaying or dying. The signatures of plankton show major alterations in species communities from 1800 (pre-industrial) to 2010, especially in the critical foraminifera spp. The phytoplankton, seagrass and related organisms are the largest group of photosynthesizers in the world, even though they occur mostly in the ocean. They provide the lion's share of oxygen to—— and the CO2 removal—— from the atmosphere. Along with forests, grasses and soil, they pretty much constitute the only really significant means of disposing of excess CO2. Ocean plankton numbers have fallen by over 40% since 1950.

Downward spiral by habitat destruction, the spread of invasive species (often carried by man), pollution, overharvesting, acidification, and eutrophication are the main factors for the degradation of waters, as well as on land. In all of the miles of the great rivers in China, 80% are now almost free of fish due to pollution and habitat destruction. Almost all of the coastlines of the world, which contribute up to 95 % of the usual fish catch, have been badly degraded. Coastline and coral reefs are the densest and most diverse main production areas of the ocean, whereas the open ocean is closer to a kind of desert; therefore, effects on the near shore are doubly or triply damaging. Significantly, 20% of coral reefs are dead, and 50% are severely threatened (Carpenter et al. 2008).

About 100 million tons of marine fish are harvested annually worldwide. (This comprises about 75% of all fish harvested.) All experts agree that the ocean fishery is collapsing. Significantly, over 95% of these marine fish have traditionally been caught in the more fertile coastal areas. But this is where the worst depletion has occurred and where climate change will initially have its most devastating effects. Moreover, hundreds of millions of people live within 100 miles of a coastline and, of course, will be catastrophically affected. Chivian & Bernstein (2018) estimated that over 50% of the population would occupy that space within a few decades.

Overfishing is the single most serious direct threat to marine animal biodiversity. The cod numbers are down by 90% (Clover, 2006). Large numbers of the large sport and meat species, like tuna, swordfish, sharks, whales and fish (i.e. cod, halibut, flounder, red snapper, tuna, shad, and similar) have fallen nearly 90% since 1950. Most devastating for the traditional fisheries, cod along the coastlines is down 99% (Myers & Worm, 2003).

The deep-sea fish, too, are being severely depleted already, even though they have only been 'hunted' in great volume by high tech for the past 25 years. Increasingly huge and technically advanced trawlers with miles and miles of nets dragging the ocean bottom have already taken a toll on deep-sea populations. The North Sea blue whiting and orange roughy, two of the major deep-sea food fish, are down to 25 percent of the numbers in the early 1990s (Clover, 2006). As noted, populations of most of the main game fish, like tuna, cod, shad and many others, have crashed even in the open ocean. The iconic cold-blooded, pelagic carnivores of the sea, including sharks, rays, skates (and even whales), are down in numbers across the board, and many reduced up to 90+ percent. A UNFAO assessment in 2005 concluded that at least 75% of the world's stock of fish was either: "fully exploited", "overexploited", or "significantly depleted".

Stark words! Clover says this annihilation-in-progress is wiping the ocean almost clean of the top of the food chain inhabitants. What the result of that eventually will be to the rest of the ecosystem is in little doubt. The oceanographers K. Jones and James Watson reported that only 13% of the open ocean has not received significant alterations by man. Due largely to depleted stocks and more difficulty in catching the most desired species, the total harvest has been sliding pretty steadily for the past 20 years. This is despite more strenuous efforts being expended by larger vessels, more open ocean netting, more technology, etc. Besides overfishing, the downward spiral of habitat destruction, the spread of invasive species, silting, warming, acidification and pollution are the overwhelming other factors in the degradation of the whole marine ecosystem.

"Who owns the ocean?" is an unanswered existential question. Currently, it is the fishermen of the world who seem, de facto, to own it. Clover (2006) details how the worldwide fishing industry operates. He says that the fishing industry is the size of the world's lawn mower industry. He asks whether people would be as willing to turn over control of lawns to that industry as they have been willing to essentially allow the fishing industry to the ocean. The sea is an almost textbook example of the tragedy of the commons; it is not owned or regulated by anybody in particular, so nations, companies, etc., each individually see value in getting as much as they can as fast as they can. Overfishing and overharvesting is the obvious result.

The effect of rising temperatures and carbon dioxide, of course, is currently the major killer threat. Due to warming alone, the oxygen levels in the top layer of the ocean have fallen 2-3 percent, which is placing additional stress on coastal fauna and fisheries. Kolbert (2018) summarizes some of the essential features of the devastating effect of excess CO2 on the oceans and their biota. Since the Industrial Revolution, man has sent over 354 billion tons of carbon into the atmosphere from fossil fuels. Deforestation "contributed" to another 180-billion-ton increase. (Much of this happens because those missing trees would have sequestered 180 billion tons of CO_2.)

More CO_2 from the air now enters the water, too, than ever (in the past 60 million years at any rate). CO_2 levels were always pretty much in balance, except for occasional spells of increased or decreased amounts due to heroic volcanic activity, tectonic upheavals, meteorite strikes, fires, or other very rare, large geological events. It had been more or less stable for millions of years, but now much more is going in than comes back out. Prior to 1800, the CO_2 in the atmosphere hovered around 280 ppm, but by 1910, it was already around 300 ppm. By 1979, it was 330 ppm and 386 in 2009. In 2021, the level, still rising, was 417 ppm, the highest in at least 60 million years.) Again, it should be noted that ocean chemistry, in general, is, by far, the most important factor in regulating the status of the air and water vapor, which essentially means all life on Earth.

As a result of higher CO_2 levels, the pH of the ocean has dropped from the long-time historic value of 8.2 to 8.1 now. The pH is expected to fall to 8.0 by 2050 and 7.8 by 2100, with disastrous results for oceans. This is because the pH strongly affects the internal biochemistry of all species, especially marine ones. This also changes the makeup of microbial communities, with more far-reaching effects.

One of these effects is that it promotes anomalous spikes in toxic algae. Changes in pH are especially bad for the "calcifiers" of the world, such as starfish, clams, oysters, sea urchins, foraminifera, sea snails, brachiopods, corals and hundreds of others. These essential species normally use ionic calcium carbonate to build their shells. Lower pH converts ionic calcium carbonate to molecular, uncharged $CaCO_3$, which is unusable for most species.

This effect of CO_2 acidification on corals and reefs has been the major reason why coral reefs have diminished in size by 50% in just the last 30 years. Coral reefs are the oldest and largest 'construction enterprises' in the world. They are the backbone of the richest ecosystem of flora and fauna anywhere in the world, supporting a million or more species, most of which, therefore, also face great peril.

Another factor affecting coral reefs directly is the rising ocean temperature. This causes coral bleaching ("white reefs"), which stops growth. Widespread bleaching occurred in 1998, 2005, 2010 and again in 2019 and continuing. In addition, overfishing contributes to occasional toxic algae blooms, which damage corals. In places, as will be noted later, agriculture runoff also greatly increases algal growth.

Thus, nearly a third of coral species are either extinct or facing the high danger of extinction (Hughes et al.). If nothing changes, all coral reefs will quit growing and begin to die off. That, of course, will lead to massive alterations and die-offs in the incredibly rich adjacent flora and fauna that depend on them. In August 2019, the Great Barrier Reef Marine Park Authority issued its latest "Outlook" Report. They rated it "Very Poor". (Their last report, five years ago, was "Poor".)

Of course, there is another scary fact of life about the ocean: it is rising rather fast and will, within this century, create more havoc for more people than any other natural or unnatural phenomenon in the whole history of H*omo sapiens.* According to Lester Brown (2012), the sea level has risen 3 to 6 inches since 1993 (average rise, 3 inches). It is projected to rise 19 inches by 2050 and about 3.5 -4 feet by 2100. It will keep rising to 9 feet or more eventually. If (when) Greenland melts completely, the rise will be 23 feet, and if the Antarctic ice sheet breaks up (as it is already beginning to do), the sea will rise 39 feet (Young & Pilkey, 2009). Most of this *is* due to global warming.

Unfortunately, the last several years have seen Arctic temperatures rise twice as fast as the rest of the globe. In 2021, a temp of 100 degrees was recorded, the hottest ever in the high Arctic. Mark Serrage, Director of the National Snow & Ice Data Center at the

University of Colorado, stated that the extra heat and loss of snow and ice are producing "Arctic Amplification". This leads to even faster melting of the ice now. As noted later, the ocean is the largest storehouse of methane. This is stored frozen on the floor as methane clathrate, so if this melt, which is now likely, global warming will become unstoppable. [By the way, a new world record temperature was recently set on July 5, 2023. The world's mean temperature was 62.92 degrees Fahrenheit. For perspective, back in the 1970's and earlier, this number hovered around 61 degrees or less.]

An obvious problem that must be seriously dealt with soon is the fact that nearly a billion people live within 100 miles of the present shorelines. David Wallace-Well's 2019 book documents in terrorizing detail how bad life will become for at least a third of the world in the next few decades. Along with sea level rise, wildfires, wars, economic collapses, and millions of refugees, he paints a dismal picture. The displacement of people and the loss of property and livelihood, which are beginning already in many places, will be horrendous. (Miami and New York, are you listening?)

In freshwater habitats, too, there is a marked decline in species and numbers of normal fauna, including most especially frogs, turtles and salamanders. There are also 364 species of endangered freshwater fish in the U.S. The Living Planet Index organization reported that their index for all vertebrates fell 40% overall, but the Fresh Water index fell 50% from 1970 to 2000. Irrigation, dams and massive engineering projects around the world have, within the last 75 years, been a major cause of extinction of hundreds of aquatic animals. For example, dozens of Tennessee Valley snail darters, snails and fish species have gone extinct within the last 75 years, mostly due to these large reclamation projects. Over 80% of the world's wetlands, too, have already been lost in the past 100 years. In the U.S., it is 85%. The loss of these habitats alone has been responsible for the extinction of hundreds of their former inhabitants.

In this chapter, we have outlined the principal effects of the all-too-many men's assault on the natural world, which we are just now beginning to grapple semi-seriously with. We have seen the explosive trajectory of the number of people inhabiting the earth. We have asked what is the current status of the <u>effects</u> of all these men on the biosphere, the atmosphere, the lithosphere and the hydrosphere.

The short answer, unsurprisingly, was: Horrendous! All is not well in our civilized Paradise. The planet of the apes has been conquered; conquered, paved and enslaved. Man is adversely affecting virtually every aspect of almost all of Nature's things as they interact with the animate and inanimate world in its inscrutable machinations. We are, in fact, heading for ecosystem damages and dysfunctions and collapses, or near collapses, on the scale of the previous five major extinction events, where the number of species dying amounted to over 50% to near 90%. Those were all catastrophes beyond comprehension, although, of course, man, or even most mammals, had not yet appeared on the earth for any of these prior extinctions. Also, all but one of the previous extinction events played out in a very slow time: millions of years. Now we're looking at damage and extinctions happening in just decades—and this time, most likely getting us too.

"What's the fuss about?" you may ask. "I don't see bodies lying out in the street, no slime hangs from our trees, no death and destruction obvious on my street, my town, county. The fields and trees are still green, with many beautiful vistas. Most people I know go about their daily lives normally and quite happily. We are wealthier and (arguably) healthier than ever. How can it be that people can warn that great and terrible changes are about to occur on a local and worldwide basis? Troublemakers! Who cares if some insects, bugs, lichens, toads and birds are getting rare?"

Even if you don't care, there is still good reason to worry. We also have a very good reason to act. We have reached the end of a great era, the greatest millennium in the history of man. But now we have a grizzly on our tail. World warming, massive extinctions, food, water, and air ecosystems are dangerously out of whack and heading badly. Look behind you; a grizzly really is on your tail. At the least, there will be a grisly train wreck in our future, partly due to rolling natural disasters like climate change and partly from the resulting human strife, turmoil and violent upheavals.

We wish that we were wrong about our assessment, which is not sanguine, about the future. In this century or next, unless energetic and rational actions are taken very soon, unprecedented calamities will befall our own species as well as so many others. (If this happens, maybe you can find some ironic solace in the fact that the earth and at least a fair number of other living things, including many plants, small animals and the miniature and microscopic worlds, will at least continue.) As EO Wilson sardonically put it:

"at least there will be bacteria and fungi — and probably vultures left."

From this nucleus, the next growth and evolution period of the earth, as well as the new biota, would resume, albeit from a diminished base. Small comfort.

Chapter 4
Biodiversity, Ecosystems & Humans

Even a bit before the sixth extinction, currently underway, was recognized (around 1980), biologists such as Paul Sears, Aldo Leopold, Garret Hardin and others even earlier had sounded the alarm about the increasing negative impacts on nature from overpopulation. "Ike" (President Eisenhower) said in 1958, "The population explosion is the most critical problem in the world."

The UN report ("First Comprehensive Report on Biodiversity" (IBPBES), Intergovernmental Science-Policy Platform on Biodiversity and Ecosystems Services", delineated the scary state of the world's land, air and water. It summarized decades of work by scientists all over the world detailing the current status of all the major biomes. The picture, not surprisingly to any biologist or scientist at least, is not a pretty one. It should be posted in a million prominent places as an introduction to the nature and scope of the dire problem. The report summarized the myriad of ways humans are reducing biodiversity and menacing biospheric integrity. Although they were not very sanguine, the report also noted that there is still time to make the transformative changes necessary in our production and consumption of food, fiber and water. One of the guiding principles they suggest as a leverage point is "a new vision of a good life."

The "Millennium Ecosystem Assessment" group, too, recently made some startling new statements: "We have changed our planet's ecosystems more rapidly and extensively in the past 50 years than in all our history. Most of these were made to meet rapidly growing demand for food, water, timber, fiber, and fuel. While these changes have clearly contributed to substantial gains in our well-being and economic development, many, however, have benefited little. Altogether, the massiveness of the changes has produced a substantial and largely irreversible loss of diversity of life and degraded ecosystems, which imperils us all."

Parenthetically, we should note that human-caused pandemics, wars, famine, genocide, etc., in the past centuries have previously been relatively short-lived. They mostly affected humans, too, often only temporarily, unlike ecosystem damages, which are not temporary nor short-lived and which affect all living things. Furthermore, environmental disasters are increasingly the cause, not the result of wars and famines. Lester Brown, 2012, details the present and coming catastrophes waiting to happen with environmental refugees. The issue of Central and South American refugees and immigration in the U.S., of course, is already constantly boiling over. Obviously, the root cause of this immigration (which will keep getting worse) is overcrowding, food insecurity and chaos in their societies. A recent UN report stated that in early 2021, there was a record number (81,000,000) of refugees. So again, we would advocate "start solving the problems at the source."

A. More Bad News

Broken-up landscapes, sprawl, roads and industrial development in remote areas are endemic to our world. A few estates or homes, along with a few roads and fences, do vast damage and break up even a good-sized wildland ecosystem. As populations grow, with all its juggling and jostling of Nature virtually in sync with it, so goes biodiversity——down! The World Wildlife Fund noted, for example, that since 1970, the populations of wild animals as a group have declined by 60%.

Even the sounds of industrial civilization can take a toll. Sound as a pollutant has received little attention, but in a few special cases, it could be a considerable factor in animal populations. Whales, for example, are an 'acoustic species' and depend on long-ranging sound waves for navigation as well as long-distance communications. It has been reported in many cases that some of the loud noises from oil drilling and similar explosive work have had drastic effects on some individuals and may have led to some of the unexplained beaching and other disastrous events.

An interesting little note of subtle, insidious acoustic effects was recently reported by a group of scientists from Boise State University. Dr. Jesse Barber and his group recorded road noise from a rural, rather low-speed, mainly passenger car traffic. Then, they placed a series of speakers deep in a roadless wilderness area on a row of trees about half a mile long. At various intervals, night or day, they played the road noise and observed the effect on the bird populations.

Even though the traffic noise was actually not very loud, at least by most metropolitan standards, they found surprisingly large effects on some species. Some resident species left the area altogether, apparently permanently. The number of birds found in the area at various intervals after playing the tapes was found to have declined by about 28%. In some of the species, MacGillivray's warbler, for one, at the end of the summer (prior to migration), the body weights were depressed compared to normal.

Changes in land use, though, have had the largest single effect on ecosystems. Especially agriculture, which, perhaps surprisingly, is the most extreme form of human-caused dystrophy. Over 50% of all habitable land of the earth is claimed by man for agriculture. Lowdermilk, already in 1935, stated, "Industrialization and agriculture are taking big bites from nature, while also poisoning, altering, intervention at every turn, scalping, and comprehensively affecting, accidentally, the whole web of life. Soil members are the first to get hit by agriculture, especially with erosion, herbicides, and soil laid bare."

In addition, expansion of cities, burgeoning populations, with their infrastructures taking up land, habitat destruction, overfishing, climate change and similar are all now major causes of biodiversity loss. Unfortunately, the fastest new habitat destruction is occurring mostly in developing countries, where the individual ecological footprint may be low, but the 'feetprint' due to excess population is very high. Weeden & Palomba (2012) say that 95% of future population growth will take place in these very places. Mazur (2009) cites Ghana as an example of how this is unfolding. Ghana (about the size of Oregon) had a population of 7 million people in 1960. Today, it is 25 million and growing. Largely within this time frame, the country has lost (not surprisingly) 90% of its rainforest. Half of this loss has come from people having to move into the periphery of the forest and cut down trees to make room for their subsistence agriculture. It is also noteworthy that 89% of Ghanaians use wood for cooking, etc. The other forest losses come from international logging, encouraged by the Ghana government to raise State revenues.

Actually, all of the West African virgin rainforests have been decreased by about 80% over the past few decades. The biggest recent factor, especially in the Ivory Coast, Ghana and Mali, has been the increase in cocoa bean production for chocolate. To expand production, large international corporations have hired natives to chop down huge swaths of the original trees and planted cocoa groves. It is expected that this cocoa tree expansion will take virtually All of the Ivory Coast's rainforest and much of Ghana's and Mali's, too, within 30 years. To add insult to international injury, a great many of the workers in these new 'fields' are conscripted, i.e., very low-paid young workers, many of them children, slaves, really. (Of course, the population in all these countries has been exploding too.) Weisman (2017) graphically describes similar conditions happening now in a great many other countries, such as Nepal and Uganda.

The worst year ever for loss in the Amazon rainforest, too, was in 2019. Climate change is now becoming a major exacerbator of all these forest changes, along with all of the other types of pollution, e.g. soil degradation, desertification, acidification etc. These kinds of increasingly severe effects are sneaking up and stalking all around the developed world, too.

In the U.S., there has been a "modest" 25% decline in the total amount of forested or wooded land. In 1630, according to the US. Forest Service Fact & History Report 2014, forests covered about 1.1 billion acres. The current figure is 750 million acres. The forests of New England, which covered over half of its territory, were originally considerably destroyed by the original colonists. However, over the next 150 years, they were fairly effectively reforested by conservation efforts. Recently, however, the Harvard Gazette, Sept. 2107, reported that for the past dozen or more years, the new forests have been cut down at the rate of 65 acres per day. The forests, especially the rainforests, are rapidly diminishing all over the world.

Around the world, 23% of grassland, too, has been degraded, and this is increasing by 2500 square kilometers per year. Grassland ecosystems, in fact, are the world's newest endangered type. Desertification is the worst of the ecological effects of grassland destruction. The largest of these endangered areas are now in Tibet and China, mostly due to overgrazing, climate change, and pollution. The Gobi Desert in China is fast advancing southward. Dunes are building only 45 miles from Beijing. Sub-Saharan Africa poses similar huge problems of rapidly increasing desertification, with concomitant loss of cropland, etc.

Actually, although it has not yet been much noted or studied, the largest single area of new land degradation going on presently may be what is being done in Canada. The Alberta tar sands are the world's third-largest oil deposit, with at least 1.8 trillion barrels. It sits in the boreal forest and, has an area of 14,000 km^2, and is being very vigorously exploited. It is now the largest single industrial mechanical project in the world, eating up large acreages of the forest along the way.

Why is all this pawing and gnawing so hard on Nature in both the short and long term? Life invented the ecosystem system on Earth, but it had taken over a billion years before even the first primitive system of life arose. These microscopic organisms, very local and weakly-interacting life forms with limited mobility, were accidentally created first in the waters over millions of years. Some of these microbe's descendants eventually developed a system of getting their energy from the sun and their food from the earth. Later, others invented the luxury of eating some of the photosynthesizers for their food and energy.

Thus, they all depended on each other almost from the beginning. Just 500 or so million years ago, life first began establishing itself on land. Millions of new niches became inhabited by swarms of specialists, but all were part of the food chain. In it, each species doesn't live alone but in an outstandingly complex niche: food, prey etc. The binding tie for all is the food web. (We might well imagine now, however, that our biological fellow travelers, like birches, buffaloberry bushes, bees, beetles, beavers, bears or badgers, must view us as a useless nuisance— actually, an existential threat!)

Mostly, though, we don't usually even really see or note the earth's biota. Part of the problem is that most of them are microbes or small invertebrates. These are the 'little people', i.e., insects, fungi, worms, protozoa, bacteria, algae, lichens, earthworms, isopods, nematodes, mites, millipedes, ants, spiders and more. Any square foot of soil contains trillions of organisms and at least 10,000 different species, with many more unknown. These are the soldiers of the biosphere that (largely) we do not see but do need nonetheless.

Bacteria alone, in any square foot of soil, number in the quadrillions. They, and earthworms, especially, are about the most indispensable fellows imaginable. In the soil, they, plus a multitude of other little creatures, help the plants turn earth, minerals and air into food with the help of the sun. They are the ones who help create and deposit minerals, break down and secrete organic chemicals, provide food for plants, and constantly building up the riches of the soil and tilth. These are really the pillars on which the rest of any terrestrial ecosystem depends. They provide free, indispensable purification, mineralization, cycling, decomposition, water retention, and similar ecological services. The 'little people' make the ecosystem go round. Without a relatively healthy and diverse population of these wee biota, the soil ecosystem simply does not 'work' at all well, with predictable results for farmers. Hawkens flatly stated, "The world cannot be fed unless the soil is fed."

Cavicciola et al. 2019; Zhang et al. 2017, Utuk & Daniel 2015, among many others, have studied 'the little people' in ecosystems around the world. They note that 40% of the world's agricultural soil (plus their biota) is seriously degraded. Mostly caused, they say, by long-time intensive usage, erosion and ag chemical overuse. Zhang et al. (2019) say, "The microbial community, composition, structure and functional potentials are significantly altered in these degraded soils." They estimated the (hidden) cost of this in the U.S. alone, from erosion, increased pesticide and fertilizer use, etc., amounts to $40 billion.

Durward Allen said in 1954: "…All life is rooted in the soil. The earth and its vegetation are the foundation for the pyramid of life."

B. How and Why Are Ecosystems Being Degraded?

Why are ecosystems collapsing? And will ours, if current actions continue unabated? Believe it or not, ecosystem collapse can and does happen, always hidden and in slow motion. Major ecosystem collapse or malfunction is not due to just one thing, by this and that habitat loss or damage. Nor by land chopped up, or pollution itself, or epizootics and the like by themselves. If natural calamities are added on, or a particularly invasive species, or wars, epidemics, social disintegration or other turmoils are added on, the likelihood of brutal, far-reaching and forever damage becomes ever more likely.

Why and how all these ecosystems have come to be so diseased and disturbed are fairly well known. Most biologists apply the acronym HIPPO to summarize the Big Five factors that are responsible for the extinction and environmental degradation. (Again, these are the tools that Kolbert says, "We humans use to transform the world's ecological landscape".)

>Number one is **H** (habitat loss)
>Second is **I**nvasive species
>Third is **P**ollution,
>Four is **P**opulation
>Five is **O**verharvesting

    The first and foremost cause of extinction and damage is habitat loss. This leads to or contributes to the most pervasive and destructive aspects of our attack on Nature. Most of it is also, pure and simple, the result of human activity——too much and too many. In some parts of the world, e.g., Madagascar, Polynesia, Philippines and others, (all of which had originally very rich species count), 90% of the natural habitat is already gone.
    One sorry example of species decline with habitat destruction is happening now right next door. Since 2001, the number of Canadian caribou (reindeer) has dropped 99% in the biggest eastern herd and close to the same in the big British Columbia herds. The Wildlife Conservation Society of Canada is worrying that they may be headed for a quick extinction in the wild. All this, they ascribe to habitat destruction by cities, towns, road and rail construction, oil and gas expansion, agriculture and the usual results of growth in virtually everything.
    Deforestation has been, and still is increasingly, a major factor in biospheric degradation. It is occurring breathtakingly fast in many places, including our precious rainforests. We have already made reference to a few examples——and there are yet more to come.

    Water shortages and desertification have not received the public attention they deserve. That public notice, however, will soon come rolling in. It will become noticed because it will become inescapable for all the human trouble and agony that its' effects will plainly visit on so many places around the world. The UN Environmental Program (UNEP) estimated, already in 1989, that 60% of the world's 8.2 billion acres of arid or semi-arid land was badly affected by desertification. Deserts are growing every year by 15 million acres. They stated that "water stress is already happening in 80 developing countries, containing 40% of the world's population." Water use has doubled twice in the twentieth century and will then double again. Another warning shot!

These examples are just a few of thousands of examples of the fixed and recurrent ravages of ecosystems from habitat loss as well as invasive species and other factors. We will not go into detail here about the rest of this Hippo family (e.g., Invasives, Pollution, Population and Overharvesting), although these, too, are indeed ravaging our ecosystems every day. As one example, the Fish & Wildlife Service in 2020 moved to place the Whitebark pine on the Endangered Species List. This unique tree is found only in very high altitudes in western America and is a crucial food resource for the (endangered) grizzly bear. Most of the whitebark decline, however, is thought to be due to climate change, helped along by invasive species, like the bark beetle.

C. Ecosystems Services

Like it or not, know it or not, but we need to start paying more attention to the real world we're living in. Naked Nature is real; ecosystems are real. They are alive. They need respect. We need to stop totally transforming them by our blind, automaton-like day-to-day, head-down lifestyle.

Chivian and Bernstein succinctly summarized the "Ecosystem Services" that nature provides for free:

1) First of all, it provides the only Primary Productivity organisms for the whole earth, specifically providing food, fiber, fuel, medicines, oxygen, habitat, etc.

2) Vital services; cleaning air, purifying water, mitigating floods, erosion, detoxification of soil, nutrient cycling, modifying climate, pollination, pest and pathogen protection, etc.

3) Aesthetic, mental and intellectual stimulation; provide a sense of place and beauty, a place for escape.

Chivian & Bernstein state that the value of ecosystems, if honestly accounted, would amount to 125 trillion dollars per year. For his entire history, until "history" began, man essentially foraged off the land and absolutely depended on his local ecosystem to furnish everything he needed. He was a part of his natural surroundings——and he knew it.

Now, man is thinking that he can feel free to not only gouge life's means but endless wealth besides from the land by his own inventiveness. Now, he has basically shed his essential connection to his evolved environment, which was shaped in and by the biosphere. He (Wm. Catton's 'Homo collosus') even thinks that he can limitlessly engineer his environment. He (we) believes we can get everything we will ever need by inventing whatever, using science, technology, skill, gigantic engineering, etc. etc.— No limits! Cornucopia is here forever.

This is not to say that science and technology should not be used in the struggle ahead and for all time. It's just that it doesn't seem to us, nor most ecologists, conservationists, and others, that scientific magic bullets are always available for us. There are, of course, many valuable suggestions from scientists and others for helping fix or ameliorate some of the worst of our biosphere challenges. One technical way suggested, for example, is to develop large-scale sequestration of CO_2, which would mitigate the major driver of global warming. The scale, cost and difficulty of doing all this, however, has been widely noted (and terminally scorched by Wallace-Wells, 2019).

Another possibility that is being discussed in various places is a large expansion of nuclear energy to eliminate virtually all use of fossil fuels. Large-scale preservation of wild areas to save all of the threatened and endangered species is seen as another possible way. Certainly, something like Wilson's 'Half-Earth' proposal is well worth starting to do, even while the larger problems are being worked on.

Some of these scientific-fix-type propositions have been referred to, and sometimes derided by various writers (e.g. Brennan et al. 2019), as the "Noah's Ark method of salvation." This referred initially to the idea of setting aside portions of biota and areas so that at least many can survive, at least minimally. The term also, however, implies that man, and only man, can simply pluck the problem up and apply the fix. Many of these proposed fixes, though, might better be seen to be cornucopian pipe-dreams, mere band-aids, i.e. increasing the dose of medicine that is already making you sick. They also seem to not take into full account the energy and water balance realities of the planet and the effects of our activities on the total ecosystems. It also minimizes the central fact: that the heart of the problem is that man is overwhelming and damaging too- large a part of nature, plundering it really. He needs now to become a lead author of the solution in getting us out of his generated mess. As Heinberg (2011) and others emphasize, to survive for long, a society must (obviously?) live within the earth's physical budget and only use sustainable physical resources. In plain English, 'we cannot consume more than the earth can replace.'

So, if people are expecting science to save us, forget it. In the first place, it is science and technology that has allowed our species to overharvest and over-run our land, air and seas. This has led directly and indirectly to the health and ecosystem diseases and disasters we now have. The coal- clogged air of Dickens over London and its lethal fogs and the rising of the seas now with wholesale extinctions offers proof enough. The high- tech irrigation systems in our High Plains did allow the Dust Bowl to become the "breadbasket of the world" —— temporarily!

Of course, we should continue to use the enormously useful methods, findings and explorations of all the sciences to improve human understanding and well-being in virtually all of our future endeavors. It's just that there are no magic bullets to pull us out of any or all follies.

Daniel Goleman (2009), in his entertaining book "Ecological Intelligence", gives some interesting insight into how humans are actually interacting (like in any ecosystem) with the environment through their industries, food, daily activities, entertainment, shopping, etc. The main point Goleman makes is that our Ecological Intelligence is quite limited. He emphasizes that we make most decisions based on very little factual verity or reasoning but instead on what he calls the 'vital lie'——"what you don't see or know, doesn't matter' [to you]. He therefore urges 'training and knowledge' to be used by our brains to train it to "see" Nature. Then, we could more quickly evolve culturally and apply technology and other information more effectively. Thus, for example, since global warming occurs slowly, we don't see it, or at least it makes a very minimal impact on our overall assessment of things. Charles Mann (2018) also talks about short-sightedness. He says that since,

"Climate change won't raise sea levels to catastrophic levels until 2100; it is too far away, so nobody cares. Even a headline like 'All 7.8 billion of us will vanish within decades' wouldn't bother us that much either."

D. So Where Do We Find Ourselves Now?

The new geological epoch we have entered is widely considered to be the Anthropocene. Where man is the central— and worse, the conquering and controlling force over the entire earth. Actually, we would suggest that we have gone even further— we are now in the epoch of the "Humanosphere." Mishra, however, (2017) simply calls our age "The Age of Anger."

E.O. Wilson gave a particularly brutal, biting excoriation of the past and present status of our human story and of people who were "Too greedy, shortsighted and divided into warring tribes to make wise, long- term decisions or plans. There has been no greatness in understanding human evolution and how the biosphere gave rise to mind, mind to culture, and culture to our present fix, where we're playing a false God, willy-nilly destroying the earth, our living environment and are *pleased with what we have wrought."*

Cafaro (2012) puts it succinctly and eloquently;

"Humans need commitment to an end to conventional economic and demographic growth. Humanity can and should continue to grow morally, intellectually, spiritually, creatively and grow in understanding, including for the first time, understanding nature and its, and our, legitimate places."

Durward Allen saw things clearly back in 1954. "Human culture is now considered to actually be a 'geological' force. But if we are at a stage where our actions are to decide the world's future, then surely we have also reached the point where we can also be held accountable for the world's future. The nobility of man would be a vain and farcical idea if the earth is to be parceled out until every individual is competing with all his equals for a meager share of pure air, clean water, green grass, and cool woodlands."

Another product (the latest and greatest, actually) of cultural evolution is moral reasoning. This innate ability, along with a neurobiological understanding of our nature and education, should become aids to advance cultural evolution. A new brand of enlightened self-interest, along with a completely revamped politico-economic system, has been called for by many people recently. Especially Terry Glavin, EO Wilson, Herman Daly, Tim Jackson,

Jeffrey Sachs, Blewett & Cunningham, Rutland Bregman, GB Frey, S. Pinker, Naomi Klein (2014), Pahl (2007), Misra (2017), Andersen (2020), McKibben (2019), Diamond (2012), Parker Palmer, 2011, and many others have talked about some reasonable ways to modify our economics and governmental approaches, as well as our social structures. Especially do we need to integrate into future societies some of the aspects of history, culture and knowledge of the indigenous peoples. While much of this has been neglected and passed over, we will find there are many things we will need from them. (One good example is Robin Wall Kimmerer's 2013 book "Braided Sweetgrass". (We will discuss this aspect again in later chapters.)

Thus, it is that the biospheric damages that are being inflicted are not usually glaring or even much noticed by the average person—YET. But they are, in fact, killing people even now, and the worst is yet to come. Famines since 1968 in Ethiopia, Somalia, China, Korea, Afghanistan, Nigeria, Ghana and many others have killed over 500 million people. More than 50 countries are net importers of a significant portion of their food. Starvation in Nigeria, which has the highest birth rate (5.4%) in the world, may be on the clearest possible path to the coming apocalypse. Its current population of 152 million today is expected to hit over 400 million by 2050. (What could go wrong?) Even more depressing is that virtually all of its neighboring Saharan countries, already badly overpopulated, will double again by 2030. These countries are the ones that most desperately need the assistance with applying the 'medicine' that we explain and advocate in this book.

The past 75 years saw the worst of the ongoing and historic blitz of the natural world. This often acclaimed feat, i.e. essentially conquering every square foot of earth, was accomplished by all of us living and the several previous generations. But we, even the last three or four generations, didn't understand what we and all the older generations had been doing. They (we) thought were just going about the usual, normal human business of pursuing happiness as humans always have done the best we knew how. 'God's work'. And they (we) did a lot of good things, didn't we?

Yes, yes, we did, and are. We have blithely and overoptimistically 'tamed the wilderness and won the West'. Along the way, we have "accidentally" mostly bettered our conditions and the whole world by our actions over the last five centuries. But wrecking large parts of the world as a result? And continuing to keep wrecking? Is that what we ought to keep doing? We now know better. Don't we?

It will be well to remember, too, that it is primarily our social theories and myths and our economic system of rampaging capitalism which is now causing and also largely hampering solutions to these dire problems. Changes in this arena will have to be recruited to be a major vehicle to get us over the hump. Arguments that will be lodged against the present proposal to slowly ebb populations will be economic, social and political. They are not ecological (or even smart or rational). However, the worst outcomes of population growth are ecological and not economic.

Greg Pahl (2007) pointedly said, "Population——or more to the point, overpopulation—— is the largest problem of all and is driving most of the other difficulties we face. But since it is basically impossible to have a sensible discussion of this taboo subject, even amongst otherwise intelligent and rational people, it is increasingly unlikely that we will be able to resolve this situation voluntarily. Consequently, Mother Nature is about to resolve it for us involuntarily."

Chapter 5
Water and Air: Imminent Crises

To this point, we have outlined some of the major features of what the sheer numbers of people are doing to the biota and biosphere. We have briefly described some of the ecological harms being done to the oceans and other waters. In this chapter, we will catalogue some of the major perturbations in the world's fresh water and air.

Life began in water, and of course, all living things are just essentially membrane-bound bags of water with organized organic and inorganic molecules arranged within. Thus, it is no surprise that we have got ourselves into a jam over water. Water scarcity and water pollution are becoming ever larger over wide swaths of the world, including the western U.S. The problem is actually leading us toward a global disaster.

During the course of Man's evolution as *H. sapiens* a few hundreds of thousand or so years ago, down to about 20 thousand years ago, men were few. They were 'just one of the boys' among millions of other species. They also had, for the most part, little effect on their environment. As documented earlier, he did, though, as his numbers and social ingenuity increased, begin to have a limited effect on his surrounding biota, especially the big ones. Ones he found delicious or threatening. As we saw, these effects involved helping to exterminate a few of them.

Later, though, especially in some of the farming areas of early civilization, Man began to have more significant impacts on his immediate environment. Already, Herodotus, the ancient Greek historian, wrote that "Man stalks across the landscape, and deserts follow in his footsteps."

Indeed, the biologist David Ehrenfeld found that deserts did follow the footsteps of Greeks, Romans and others.

He said: Desert-maker is truly as appropriate a title for humans as 'tool-user."

Thus, a shaky equilibrium with nature lasted until man's numbers reached nearly half a billion six or seven centuries or so ago. Then, his numbers began to be highly significant to others of his fellow Earth travelers. With the conquering of the New World and soon the Industrial Revolution, his air, land, and energy uses and ecological clout would increasingly become the major disruptive factor in relation to the environment. In fact, his water usage and abuse alone have now led us to a very serious problem and a crossroads. The rapidly increasing number and scope of larger and larger metropolitan, agricultural and industrial practices have had profound adverse effects on drinking water, disposal systems and all aspects of water (as well as everything else in Nature). In fact, the situation is getting worse every day. Actually, it is a "hair-on-fire" emergency already that almost nobody seems to believe, and even fewer are willing to do anything about.

A. Drinking and Domestic Water

All water, of course, comes from, ultimately, the earth's basic stock as part of a water cycle. It starts from rain and snow to form fresh (and oceanic) bodies of water, rivers, lakes, etc. Some goes into the ground as groundwater. Much of the surface and subsurface water is taken up by plants (and thence much of it into animals). Water transpiration from the plants, along with evaporation, eventually returns most of the water to the atmosphere, to again fall to the earth as more precipitation, and so goes the hydrological cycle. In this whole cycle, we have long had about the same absolute quantity of water. There is little net gain or loss, and we have been mostly using the same ancient water for as long as life has been on Earth.

Nevertheless, drinking water, actually all freshwater, is today an endangered commodity, mostly due to the draining of groundwater sources and pollution of the surface, along with the growth-induced constant local demand for more and more water. Safe and abundant drinking water is the most pressing problem facing about half the people around the globe. (NB: There would, however, be plenty of good, sustainable water available for, say, half a billion people.)

- A few recent reports from various news and scientific papers summarizes the grim water prospect.

- 60% of the world's 8.25 billion acres of arid and semi-arid land is affected by desertification." (UNEP, 2017)

- The Colorado River has no more water to give to already water-starved western U.S. cropland.

- Groundwater levels are dropping fast in 18 countries (home to half of the world's population)

- Hebei Province, in the heart of China's grain belt, aquifers dropping 3 meters per year.

- Almost half of the 800 largest cities in China is likely to suffer a severe water shortage by 2040.

- In India, there is a crisis today. In the Lowlands of India, groundwater is falling even faster; already, drinking water is being trucked in.

- 33% of the world's population lacks access to clean, safe drinking water.

- The Po river in Italy today is starting to dry up.

    The National Institute for Transforming India (NITI Aayog), a group within the Indian government, recently issued a report. It starkly laid out how beginning in 2020, 21 Indian cities will start to run out of groundwater and drinking water due to drought drying of aquifers, reservoirs and rivers. Further, they state that barring decisive action and godsent heavy moisture, 40% of all India will be in imminent danger of running out of drinking water by 2030. Already, problems of shortages and unspeakable pollution are affecting the vast majority of Indians.

    Other Asian and African countries are facing similar disasters. While in most of the developed countries, the fresh and groundwater situations may be less dire and further away from disaster, it is not all that far from it either. Charles Fishman ("The Big Thirst") explores the dire water situations all around the world, from Las Vegas to Los

Angeles, Australia, India and many other places in between. Mekonnen & Hoekstra (2016) state that 4 billion people are facing severe water shortages already. This will certainly lead fairly soon to a world-shaking human tidal wave of suffering. Half of the world's largest cities experience regular water scarcity. Demand for water "will outstrip supply by 40% if nothing is done" (World Economics Forum, March 2007).

Pakistan, the whole country, has recently had its water status changed from "Water-stressed" to "Water scarce" status. One of its largest cities, Quetta, population of 1.1 million, has a water system virtually unchanged since 1951 when its population was about 84,000. The "Integrated Regional Information Networks" in Pakistan stated back in 2002 that "Quetta will likely soon become a 'dead city.'"

Plain old water pollution, from either human or natural causes, remains a worldwide problem, worst in the overcrowded or underdeveloped world but 'bad' basically all over the world. The UN and other agencies report that rare pollutants found in essentially all water throughout the world pose their own disease or other problems. Five thousand children, mostly in Africa and Asia, die of diarrhea alone every day, most caused by unsanitary water. Over the whole world, the WHO stated that nearly one-third of the population (2.5 billion people) lack access to safe drinking water. Further, the U.S. CDC projects that 1.8 billion people also do not have access to adequate sanitation. (These stats are essentially immutable in future, too, if even the current number of people is continued.)

Another kind of problem arises with competition by countries for the fickle waters of the Nile between Egypt, Ethiopia and Sudan, for example. Already, 95% of Egypt's population is facing a crisis, with improperly treated drinking water due to poor sanitation, pollution and shrinking sources. A study by a group of scientists at Dartmouth recently found that by 2030, demand for Nile water will vastly exceed supply. (Egypt, for one, receives 85% of all its water from the Nile. The other surrounding countries vying for the water are similar.) Ethiopia has recently completed the Grand Ethiopian Renaissance Dam, which is now still filling. It could cut Egypt's (already inadequate) supply by 25%. Similar competition for the Euphrates River exists between Syria, Turkey and Iraq. We have not heard the last of conflicts and catastrophes ahead soon for this region. This area will increasingly become virtually uninhabitable for a majority of their people well before the end of the century.

Beyond the problem of not enough safe drinking water in places, there is rampant pollution of freshwater from industry, irrigation runoff, etc. Water pollution occurs in all types of water. Air pollutants, too, readily enter rivers, lakes, streams and other water bodies. Inevitably, these then 'trickle down' eventually into virtually everywhere. Direct addition of pollutants from agriculture, industry, homes and even other parts of the ecosystem are also a constant source of one type or another water pollution (often very severe ones). Lake Erie, for example, has several "Dead Zones".

Agricultural runoff with fertilizers, pesticides and other chemicals is a huge problem as they work their way into virtually all waterways. In the Mississippi River drainage alone, it produces a dead zone in the Gulf of Mexico the size of New Jersey. The ocean here is basically a near-lifeless desert, and it will continue to grow. There are 500 smaller dead zones around the world.

Plastics in the water are a new and increasingly vexing problem, particularly in the ocean. A recent report by the McArthur Foundation said that at the present rate, there will be more plastic in the ocean than fish. In just the Great Lakes, 22 million pounds of plastic are dumped each year. Not only does plastic create a physical problem, like when marine mammals, whales even, get tangled up in various plastic debris, but also by mechanical intestinal blockage and other effects of ingesting them. (Not to mention the disgusting floating islands of discarded plastic that constitute probably the worst single blighted area on all the Earth.) UN data says that plastics are killing up to 1 million seabirds and 100,000 mammals and turtles annually. Recently, it was calculated that the total mass of plastic in the world is the same as all mammals combined.

Microplastic debris is an increasing problem, as new reports blossom regularly of potentially toxic microparticles in many species and humans as well. Recently, a report documented large numbers of < 5mm plastic particles in the snows of the far Arctic, apparently carried there by the winds. Even the ubiquitous plastic water bottles produce potentially damaging microparticles. More disturbing yet, birds, turtles, fishes, sea lions and even small animals eat large and small plastic residue, some containing miniparticles. These get clogged in their intestines and, in some cases, cause disease from toxicity. These microparticles are all over the ocean. Recent deep-sea sampling has shown a large number of these particles in the sludge at the bottom, too. The World Wildlife Fund (Oct 2018) laid out how plastic is a major toxin in, especially marine birds. Plastic particles, surprisingly small to surprisingly large, are found in the gut and/or tissues in 90 percent of marine birds (that number is 100% for Midway's birds, including the iconic "Goony bird" ——the albatross).

A particularly gruesome recent report, although not occurring in water, involved camels in the United Arab Emirates. In a multiyear study, the scientists studied over 30,000 dead camels in the country and found large polybezoars (compacted plastic) in the stomachs of 1 percent of them. The largest one weighed 140 pounds. These caused the camels to slowly starve.

Again, the answer to the plastic problem is potentially quite simple: don't make so many non-biodegradable plastic or similar containers (or so many plastics, period). There are several newer plastics already, like polylactics, for example, that are more easily and quickly degraded. Other, more advanced types of material are being researched. Let us hope that—— population reduction aside— a relatively quick partial technical fix can help us out of this immediate huge problem. This is one place where science could help greatly ameliorate a problem.

B. Water, Agriculture & Irrigation

Agriculture is actually the largest single contributor to water pollution, affecting the soil and water directly and the surrounding ecosystems indirectly. Here, too, the main root problem is too much land being used for too much (often huge) industrial agriculture. With the rise of giant monoculture agriculture, mostly driven, again, by population pressure, the demand for water has increased exponentially. Water for agriculture, as presently practiced, and upon which much of the world desperately depends, is already producing an enormous problem. It is the single largest usage of water almost everywhere in the world. The percent of water used in agriculture alone around the world is estimated at over 70% of the total water used. Already, a great many of the huge agricultural epicenters are unsustainable and well into the process of running disastrously low (like California and the whole desert Southwest). Absent a decrease in long-term demand (i.e. decrease in population), unspeakable disasters await here and in a great many other places.

The area irrigated in the U.S. is over 55.8 million acres. This consumes over 130 million acre-feet of water, which amounts to 47 trillion gallons per year. Unfortunately, a great deal of this water comes increasingly from groundwater, which is essentially geological water and only very slowly replenished. As a result, in the U.S. (actually, all around the world), the majority of groundwater reservoirs are being depleted at a rapid and unsustainable rate.

There are ten major underground water reservoirs or aquifers in the U.S., which supply a majority of the irrigation water. California alone uses 16% of the U.S. total irrigation water. Arkansas and Texas each take 10%, Nebraska gets 6 percent and Idaho 5 percent. Underground water, unfortunately, supplies over 50% of all the freshwater, too. The Ogallala reservoir, for example, besides the enormous irrigation enterprise, provides drinking water to 82% of the 2.3 million people who live on the High Plains.

The Central Valley aquifer in California originally had a capacity of over 200 trillion gallons, mostly used now for agriculture. It is losing around 3 trillion gallons per year. All of the California reservoirs have a total capacity of only 850 million acre-feet or 270 trillion gallons. Overall, California irrigation uses 40 million acre-feet per year (13 trillion gallons), which irrigates 5.7 million acres. Altogether, the irrigation-fed arid farms of California produce about one-third of all U.S.- grown vegetables and two-thirds of fruits and nuts. All of these aquifers, as well as the dwindling surface waters, are shrinking and severely threatened—actually a flat-out emergency today. These depleted aquifers and drought are beginning to cause unprecedented problems in agriculture. Shortages of certain foods will fairly soon start to appear locally over significant areas, even in the U.S. A wake-up call, if ever there was one, and it will only get worse in the near future.

Farmers and other users, however, are still crying the usual song–—"We need to divert other rivers and develop more ways to acquire more new supply." In California, they are currently crying, "Why are we letting so much of our river water flow out under the bridge in San Francisco and out to sea, wasted." However, it should be obvious that our present gigantic efforts to try to maintain enough water for our food and other uses are strained beyond repair and will have to be drastically cut back, not increased, or continued as-is.

The recharge fluctuations of most aquifers are not really very well determined, as year-to-year and decade-to-decade fluctuations are so wide. There is no disagreement, though, that the process is painfully slow. Snow and rainfall seep into aquifers at very slow rates, meaning that if their level falls, it takes years, if not decades or even centuries, to raise their levels significantly. The crucial Central California aquifer, for example, has been lowered more than 300 feet in the last few decades alone. At current levels of use, it will fairly soon run essentially dry. For many years, wells have continuously needed to be drilled deeper and deeper, exacerbating an already disastrous situation. By the middle of this century, the Sierra Nevada snowpack, which essentially feeds the entire valley, will fall by up to 50%. The situation is getting more dire practically by the day. ("Ho–hum?")

The prospects for irrigation-as-usual (indeed, life as usual) from all of California's water sources, actually all of the desert Southwest, are exceedingly grim. Philpott (2020) lays out the details of their plight in stark terms and predicts major disasters and agriculture collapses in the next couple of decades. All experts agree on one thing, and this goes for the Ogallala Reservoir and all the others, too. Underground reservoirs (like the food supply situation) are barely enough even for now but will run nearly dry in the foreseeable future. (One of those pesky limits!) "The accelerating depletion of aquifers means that the consequent drop of harvests will create probably unmanageable food scarcity" (Brown, 2008). There will be no engineering or technological fixes available for that. Only decreasing the demand for water can solve this problem. Current plans and outlooks, using just short-term thinking, are pathetic.

The largest aquifer (one of the largest in the world), the Ogallala Reservoir, underlies eight midwestern states. It covers 174,000 square miles under the high plains and irrigates 7 million acres. It supplies 30% of all the nation's irrigation water. According to Fishman (2011), irrigators in the past have used as much as 27 trillion gallons from it in some years. Its total storage capacity has been estimated to have initially been perhaps about 900 trillion gallons.

It generally lies 50-750 feet under the surface and, on average, is only 150-350 feet thick. John Opie, who wrote the book on the Ogallala ("Ogallala"), said that possibly one-third of its usable capacity had already been used by 1993. This aquifer, like most of them, is recharged by rain and snowmelt only in glacially slow amounts. In many cases, the rate is less than half an inch per year: in other words, not at all. (In some aquifers, though, in some favorable years, the recharge rate is much higher. Still, it is always much lower than the current use.) Virtually all of the water in the Ogallala aquifer is "fossil water", which was charged up following the last of the ice ages and is essentially unreplaceable.

The sad, unfolding history of the Ogallala reservoir (currently playing out almost in seeming slow-motion) may be a good proxy for a canary in the coal mine. Into the region first came the dryland homesteaders in the early 1890s. They did no damage to the aquifer, but they did come in droves and eventually drylanded much of the region into the Dust Bowl (see Eagan "The Worst Hard Time"). Later, the farmers turned to technology——irrigation. Irrigation, it turns out, will result in even worse times for the region, although so-far, so-good (as seen from the standpoint of people watching a man fall from the 35th story as he passed by the 30th floor).

Industrial-grade irrigation began about 1960, and by 1993, one-third of the aquifer may have already been used up. A 2017 USGS Report ("High Plains Aquifer Groundwater Levels Continue to Decline") said that the level is dropping by 1-3 feet per year. Much of it had dropped about 150 feet total since 1960 when irrigation started at a massive scale. A recent article in the Denver Post stated the Ogallala in 2018 "shrank twice as fast as it had over the past 60 years." Scientists at Kansas State University have recently reported that the aquifer will be over 70% depleted by 2060 if drastic changes are not made to usage. Usage, however, continues virtually unabated, although some small advances in efficiency have been gained. Attempts by State, Federal and local groups and agencies are actively in play to demonstrate ways to convert current industrial ag activities to a sustainable dryland situation. All the farmers, too, know (or should know) that their groundwater bonanza will run irretrievably dry in a few decades.

Nevertheless, there are still about 150,000 pumps running practically day and night for half the year. But, the water is virtually free, and most really can't see that they have any other choice, especially since few people factor- in the environmental as well as future costs. Maximum yields for maximum profits are what still mindlessly drive the enterprise. While present attempts to alleviate the problem may have resulted in some small decreases in the rate at which aquifers are being depleted, their ultimate fate is sealed if usage is not cut by much more drastic amounts. The efficiency of irrigation methods and equipment has been the main direction taken by the current efforts. But this approach really only just postpones, by a very few years or decades, the inevitable depletion. No farmer (probably) has volunteered to stop irrigating or to substantially reduce the acreage under irrigation.

It is like watching a slow-moving horror movie as this process continues inexorably to the last obtainable drop of water. Once, after the disastrous Dust Bowl, the High Plains became the Breadbasket of the World, but it will soon end in an agonizing ecological and human wreck. (Unfortunately, another, even larger, U.S. breadbasket, California, is also heading for a similar fate.)

So far, most of the large-scale solutions offered to fix the Ogallala problem have ranged from the zany to the too-little, too-late. Over the years, for example, groups have explored diverting rivers onto the High Plains for storage water. Rivers like the Mississippi, the Missouri, the Peace and other Canadian rivers. Some have even cast a jealous eye toward the Great Lakes. (Those eyes, however, were quickly poked soundly by powerful local entities.)

It seems to us that the problem is related to the same old problem––numbers. Too many acres using the one or two acre-feet necessary to raise crops in semi-arid lands. The only real solution, again, is to "reduce the amount of irrigation." That, of course, is a tall order and would hopefully be accomplished with something like part of a century's plan, say. If the population of the world were to be reduced to, first, one-quarter, then one-third, etc., eventually to well less than the maximum carrying capacity of the Earth (1 billion?), then the need for maximal farming acres is removed. The aquifers would then begin to slowly, over centuries, recharge to some new steady-state level. In any case, the irrigation acres will be reduced drastically in the not-too-distant future anyway, no matter what; the aquifers WILL run dry if nothing changes. Opie says

"business as usual" [as contrasted to his suggested switch to smaller-scale, sustainable agriculture] will continue as long as the marketplace continues to regard environmental and human costs as irrelevant. Political intervention is essential. We cannot take nature or successful agriculture for granted, as past generations have had the luxury of doing."

Running out of water for irrigation will be a huge catastrophe no matter what else is going on. If that happens soon (as it will), that will create cascading catastrophic situations for humans directly that are almost beyond comprehension. This will create food shortages, not so much in the U.S., at least not right away, but practically all around the world. In turn, that will cause cascading cataclysms in social structure, government, human health and virtually everything else. We are, in fact, in many places around the world, already staring, in just a couple of decades, directly at a rolling, worsening crisis unparalleled in human history. When the Ogallala runs dry, the farmers and inhabitants of the High Plains will suffer a mega-collapse, with unspeakable suffering of families. Not to mention the collapse of the gigantic food supplies coming out of there (along with California) to feed ourselves and the world.

The High Plains now contributes about one-third of the total U.S. food commodities. When (if) sustainable dry land agriculture is restored at a lower level, the food supply from there would, in a couple of decades, drop to much lower levels, back to its original dry state. Focusing an eye on the looming fate of the doomed High Plains farmers seems almost like a macabre act. But maybe doing that could result in some changes being made soon to help avert or minimize the certain, unprecedented calamity.

(It should be noted it is true that the entire Ogallala aquifer is not being so disastrously depleted. Most of the damage is being done in the southern half, while less is occurring in the northernmost area. It is possible that the northern part could hold up maybe even another century before the inevitable happens there, too.) [This situation, however, reminds one of where one person is sitting in the sinking stern of a boat while the guy high and dry in the bow says, "Just your half is sinking."] Actually, the whole situation well illustrates a basic Catch-22. People say that since we need to keep up maximum food production to feed the world, we have no choice but to also keep on depleting the aquifer. That, however, only (conveniently) postpones the inevitable gruesome collapse.

Fishman (2011) has detailed these kinds of problems with water all around the world. He concluded, too, that the problems are usually local but involve massive logistical, ecological and sociologic parameters just to get by. Many of the dire situations with water that he describes in Australia, or Las Vegas and California, for example, actually do involve severe shortages of water in the absolute sense. In some places, like California and Australia, trying to get water for colossal agriculture setups leads to competition for the already scarce water, which then starts to run short the people in cities and towns. As the aquifers deplete, the disastrous ending for everybody is not hard to predict.

Since the turn of the century, the venerable Rio Grande has run dry south of Albuquerque for many years. The Great Ruaha River of Tanzania has run dry for long periods for the past several years due to a rapid increase in population who are diverting large amounts of upstream water for agriculture. Even the Mississippi River fell to historic low levels in 2022, delaying and seriously affecting the vital barge traffic. Lake Chad, Africa's fourth-largest lake, is already 90% dried up and still shrinking. Absent a lessening footprint, problems like these in many areas will remain, practically speaking, insoluble.

Until both population degrowth and smaller and better ecological practices are emplaced in agriculture or have at least begun, these problems will remain insoluble. Small (or large) technical and other remedies are of some, but limited, use. But until massive efforts are made to tackle the root problem, there is a looming disaster here and around the world. Again, all of these problems stem from the same proximal cause: too large a human footprint on the environment. Actually, as the conservationist Dave Foreman (2014) puts it: "It's not the size of the human footprint, it's too many human feet. It will require modified ways of thinking, new theories and designs of human economics, social structure and governments."

Whatever plans, guidelines, proposals or whatever it is that people come up with over the next century, it will require massive changes in governance, public strategies, ideas and platforms in all aspects of society and all States starting soon. All around the world, these kinds of water issues continue to pile up disastrously. In many places, not having enough personal water available, in terms of volume alone, let alone contaminated water, is here today for more than a billion people–and ever more tragedy looms in huge swaths of the world.

"But," you say, "if it's true that around the globe there is plenty of water actually 'there', we should be able to use large-scale science and engineering tools along with genetic engineering, etc., to breed plants that conserve water to bring all the water we need to wherever it is needed."

We would answer, "Yeah, yes, we probably could." But that would not answer the basic problem that already bedevils us. The present use and abuse of land and water is already assisting in the vast and unsustainable loss of biodiversity and ecosystem damage. Adding more of that medicine to effect a cure of a likely incurable disease does not seem like a wise approach. The work we already do to get our water is very hard on our wildlife and our ecosystems, too. That is the problem to start now to fix first.

C. How Much Water Is There?

Normally, this would be a nerdy (or stupid) question. But it might be a worthwhile task to wrap our heads around this issue. The earth is covered around 75% by water. Most of the water, however, (96.5%) is in the ocean. Fresh water for human use comes from only about 2.5% of the available total on earth. That's part of the problem, although not really the main one. We should emphasize that there is no absolute shortage of water in and on the earth. Numerically, the number of water molecules has been pretty constant for millions of years. The only water we will ever have was left here, practically at the beginning of the planet. Some of that water is still in the rocks, making up the whole earth. Water has been used, cycled and recycled on the surface of the earth forever, although the cycle has been drastically modified from time to time by massive lithospheric, galactic or climatic events.

Man still cannot 'make' water, although it is possible (though completely impractical) using huge energy inputs or impossibly complex methods to 'synthesize' it. Still, even though we do not get any 'new' water, there is at least a constant, and basically seemingly quite adequate, amount of it. Thus, we will not really run out of it or use it all up (except for the aquifers). The very severe problems are mostly related to how to get enough for each locality, when and how needed. Lately, these problems are becoming basically insurmountable because of the exponential growth of people and demand and the enormous cost, economic as well as ecologic, in the huge geoengineering projects required. Each of these causes disruptions and wreckage of large sections of land. (There are, for example, over 80,000 ecologically impactful dams on the streams of the U.S.)

For the moment, though, let's concentrate our analysis on the U.S. only. The total amount of groundwater is estimated at about 35,000-45,000 trillion gallons. The total surface water in swamps, rivers, lakes, etc., comes to about another 30 quadrillions (30 x 10^{15}) gallons. (It might be noted that only a fraction of this is ordinarily readily usable. For example, in Lake Superior, almost 11 trillion gallons are lost annually by evaporation alone.) It is the groundwater, lakes and rivers which supply much of the 'normal' water that we use and drink every day.

All of this water, of course, comes from rain and snow. The average precipitation over the U.S. is 30.2 inches per year. Thus, each square meter, on average, receives about 250 gallons of precipitation. Theoretically, that means we have available, excluding groundwater, about 200 quadrillion gallons of rainfall and snow water for our use each year. Theoretically, that should be way more than enough since the total human usage in the U.S. is 'only' 133 trillion gallons per year.

According to the USGS Report of 2015, the total usage for the main categories of use per year are:

Irrigation	47 trillion gallons
Public, household, etc	15 trillion gal
Industrial (includes mining)	6.5 trillion
Thermoelectric cooling	8 trillion
Livestock	0.7 trillion
Aquaculture	2.5 trillion
Miscellaneous	23 trillion

TOTAL 133 trillion gallons

Out of these 133 trillion gallons, almost a third comes out of groundwater and not immediately directly from the annual precipitation. The rest comes out of rivers, lakes and other bodies of water. All of this, of course, comes ultimately from rain and snow. People didn't directly use up all that 133 trillion gallons either, except for the underground water, used mainly for irrigation. A good deal of the water is reused many times over in some cases. The Mississippi River alone empties 1.5 – 5 billion gallons every day into the Atlantic. Obviously, a great deal of it has been used at least once (and will be reused many times again), so it is difficult to tell exactly what percent of the rainwater was actually 'used.'

The main problem is that in many places, like the arid West, there is too little rainwater or any other permanent (reasonably close) water source. In some other places, there is too much. On average, then, overall, the U.S. has plenty of good drinking and personal water available going forward. The "average", however, doesn't apply to many large, local areas where there really isn't enough for everybody.

In a great many places, like California, the Great Plains and the whole southwest, there is not nearly enough to even continue agriculture as is when the fossil water (aquifers) fail (Peterson et al. 2016).

Already, many aquifers in many locations, especially the western and California aquifers and reservoirs, are way over-subscribed, i.e. unsustainable now. In so many places, requirements for drinking water and household use have already collided gruesomely with the requirements for agriculture. New supplies in a great many places (Las Vegas and Phoenix, the most egregious examples) are already unattainable without undertaking the most gigantic engineering and hydraulic and ecosystem-altering hydraulic projects at unsustainable and unimaginable environmental costs.

The calculations for many other countries, especially China, India and Australia, are even grimmer. China has been in the worst drought in centuries, as has Australia, parts of the Middle East and the southwestern U.S. The hydrologist Jay Famiglietti, using satellite data (Nasajpl, 2015), has shown that the aquifers underlying part of Syria and along the Tigris-Euphrates River are very close to depletion. Indeed, NASA satellite data shows that a majority of the earth's largest aquifers are being rapidly depleted and are grossly unsustainable.

In Fishman's (2011) survey of the water situation in India, the grim reality is almost beyond description. In talking about the conditions around Mumbai and Delhi, he says that the Yamuna River, which is the main tributary of the Ganges, supplying 60% of its water, is unbelievably polluted. It ceases almost to flow at its mouth. The India Centre for Science & Environment stated, "One eyedropper of Yamuna River water is enough to make 6 bathtubs filled with water unsafe to sit in." Fishman says, "The venerated Ganges River is, if conceivable, dirtier." The situation in Delhi is much the same.

China is facing perhaps the worst of future water problems. The Yangtze and many of the other major rivers are increasingly running lower and lower. The World Bank recently (2018) stated that over 90% of the country's underground water is "catastrophically contaminated." Eighty percent is "unsafe for drinking." Similarly, Fishman says that India, too, just "drifted" into a self-inflicted water crisis, the result of rather mundane, nuts and bolts, poor engineering and management from top to bottom, beginning, of course, with burgeoning populations and government fiascos. Mann (2018) says, "Corruption, inefficiency, incompetence and indifference marks government's poor record on water systems." The problems in India, indeed in many other places too, are insoluble, absent a decrease in demand.

Thus, the water situation all around the world is in a man-made mess. What will happen when the Himalayan glaciers melt makes us shudder. Since the turn of the century, on average, these glaciers have retreated 5 meters per year, losing eight billion tons of ice. And, obviously, it doesn't look any better going forward. This is another creeping disaster waiting to happen. A similar situation is occurring in the glaciers of South America. Chile's glaciers, which supply water to a majority of its people, comprise 80% of all glaciers in S. America. These glaciers are melting rapidly and are predicted to be at least half-melted by the end of this century. As luck would have it, Chile's glaciers are located in one of the world's largest mining districts (copper). Naturally, the government is afraid to do anything because of the feared severe economic catastrophes if mining, which comprises the largest single portion of its revenues, would have to be curtailed. (Of course, we should not forget Glacier National Park's near-complete loss of glaciers, either.)

Much of the ecological damage, including pollution of almost all water sources, is caused by Man trying to re-engineer large chunks of nature to supply water to the jam-packed cities and parched agriculture. Possible solutions to the latter problem involve almost impossibly large infrastructure, as well as political and economic programs. The teeming 8 billion-plus growing population makes essentially impossible the herculean effort to clean up and fix the desperately inadequate water supply and infrastructure in so many areas. Water apocalypse will probably be the first of the Four Horsemen.

Again, the question is not really whether the earth has enough water or even if there is theoretically enough to supply Las Vegas and California, for example. The insoluble problem is how to move the enormous quantities of water from one place to another without major damage to the rivers and lakes and all the other ecosystems that are already afflicted by huge engineering and technical projects.

To illustrate further, we can look at the Colorado River. The Colorado supplies essentially all of the drinking (and most of the other water uses, too) of Nevada, Arizona and Utah and a good part of California. (Fourteen percent of all food production in the U.S. comes from land rendered fertile by waters from Colorado.) This river, on average, only carries less than 15 million acre-feet (5 trillion gallons) over a year. Between Las Vegas and Los Angeles, plus parts of Arizona, Utah, and Colorado, we are currently consuming about 20 million acre-feet (Colorado River Research Group report, 2019). Thus, these 60 million-plus 'customers' consume All of the flowing river water and then some! The rest (the deficit) comes from underground water (aquifers) and (temporarily) stored water from Lake Mead and the other reservoirs. (Lake Mead and Lake Powell, however, are shrinking and are now at their lowest level since their original filling. Lake Mead, in fact, has just crossed "The Critical Threshold" level, where the flow can no longer supply current demand. A 5 percent decrease in allotment has been imposed upon the already severely distressed Phoenix area. Any further drop will make it impossible for the hydroelectric plant to produce electricity.)

Obviously, the aquifers will rather soon run dry, too, and then unspeakably bad things are in store in our backyard. Supplying enough water to the millions of people (and their agriculture) from Phoenix to Las Vegas and LA and all in between is all but impossible already. By massive systems of ditches, aqueducts, canals and all manner of huge hydroengineering, with the indispensable augmentation by underground water, they are barely getting by, even as the problem magnifies by increasing demand. Suffice it to say, the degenerating state of the California (as well as all the Southwest) water situation is almost too depressing to chronicle. A recent study at Carnegie Mellon Univ. found that central California alone is losing $3.7 billion in agriculture revenue annually due partly to salinity issues. The piper will soon have to be paid. There will (quite soon) be hell to pay as food production falls.

George Sibley has, for years, brilliantly documented this saddening tale of the whole Colorado River system. Also, Marc Reisner's marvelous, monumental book "Cadillac Desert", published originally almost 40 years ago, details the history of this and all of the American western deserts' mind-boggling, rocky, and increasingly dangerous dance with water. He predicted virtually all the water troubles that are now staring us squarely in our faces. He said, "Already soil, salt, silt and dwindling sources are creating the coming water Armageddon in our desert lands."

These current pressing problems here and at many other places around the world are basically insolvable if the present population (and therefore, "demand") continues. How to manage at the local and regional levels to provide sufficient and clean water for the present and future direly deficient cities and areas, even with gigantic and Nature-destroying engineering feats, is an increasingly horrible problem, hapless really. Actually, a continuation of the current population is already overwhelming local human and natural ecosystems. Again, we say the task is impossible to permanently solve absent population reduction measures.

Missing in all this, too, has been an appreciation of the costs and hardships to wildlife and Nature and the ecological damage that such huge interventions such as dams, diversions, pipelines, desalination, etc. inflict. Using massive industrial, technical and scientific methods and materials on gigantic and centralized scale as the solution to water (or other problems) might best be seen as just an attempted 'work-a-round.' Charles Mann calls this the "hard path" approach to problems. It is the approach of the Wizard (Norman Borlaug) in Mann's book. It has been the go-to approach usually taken. It is different from the "soft-path" approach (Gleick & Patann (2010), as described first in 1979 by Amory Lovins in regard to devising solutions to the energy problem. In any event, we clearly need a "new' path".

Don't say you haven't been warned............ Again.

D. Air

The atmosphere cannot be avoided by most of the millions of species that occupy the earth. In fact, long ago, these organisms invented most of the biosphere. Ever since, they have (a) produced the "air" as we see it today, (b) used it continuously to maintain our lives, and (c) been a large part of regulating and maintaining its peculiar cycle. When plants arose half a billion years ago, the oxygen level rose, and the carbon dioxide level eventually reached a more-or-less long-term balance. Handily, plants made oxygen and biomass for all the animals and 'little people' and everyone else, using CO_2, soil and water. These animals and other 'people', of course, use this food and, in turn, return the CO_2 back to the atmosphere. Some of the little people also regulate the nitrogen in the atmosphere by fixing it into usable minerals for plants. Plants are also a major factor in regulating the water vapor in the atmosphere by their vast operation of transpiration. Ordinary evaporation also contributes, in stochastic ways, to maintaining equilibrium in the hydrological cycle.

Until lately, this ageless cyclic pattern held relatively steady (albeit not without considerable, often violent, variations over geologic time periods). Gigantic volcanoes, continental shifts, meteorite strikes, extinction events (five so far) and other catastrophic planetary upheavals considerably altered the content and cyclicity of the atmosphere–lithosphere–biosphere team. The results of each of these upheavals at times created great stress for all of the biota on the earth at the time. At present, however, extinctions, for the first time, are being caused mainly by a single biological member. None of the previous five extinctions were caused, or even hardly affected by any kind of dominant animals, as only the smallest, primitive mammals had even arrived on earth by the time of the previous extinctions.

Thus, for millions of years past, our atmosphere has more or less been in rough, though cyclic, equilibrium. The maintenance of our atmosphere, in fact, is one of the services that our biosphere has provided free of charge for uncounted millennia. The large oxygen supply originated mostly from photosynthesizing plants, which, of course, continues today. A large percentage of this present oxygen comes from the ocean via planktonic algae and other green plants. All of the other plants worldwide also contribute their share. Plants (along with a host of various other microbes and similar 'wee ones') also serve us free of charge by providing us with usable nitrogenous, phosphorus and many other minerals as well as most of our carbonaceous food. They also filter out a good deal of our atmospheric 'extras', i.e. pollutants.

All this, of course, is now thrown seriously out of whack. Today, the human team is affecting the atmosphere in massive and unsettling ways. Ways that are causing major changes in the whole biosphere. Ways that we don't like. Ways that threaten our acculturated life, if not our very lives. Briefly, we will talk about two major changes we are causing just in the air——pollution and altered gases.

A. Pollution

Pollution of the air by itself is a very big deal. The problem, though, seems to always be ignored. All the excess stuff that our activities try to store in the commons that we call our atmosphere is having an effect. The list of pollutants is all quite familiar, as they have been accumulating for generations. Some of the big ones are: gases, 'fine particulates' (such as in smoke and industrial compounds), dust, toxins, chemicals, etc. All these are very big problems that almost everybody has seemingly forgotten. Nobody gives it a second thought anymore. "You don't see young or middle-aged people just keeling over and dying from air pollution." Nevertheless, every clinic and hospital in the world sees its insidious and deadly effects every day. Nitrogen dioxide gas from trucks, urban and other industrial sources, for example, is a major air pollutant and directly adversely affects human health. Worldwide, more deaths are attributable or relatable to bad air than to automobile accidents. In 2019, there were 55.3 million total deaths worldwide. Of this total, 4.6 million deaths were attributable to or related to air pollution—almost 10%. Until the last couple of years, the air quality overall in the U.S. had been slightly improving; it has been getting worse again lately (EPA report, 2019).

It has also been getting much worse all over the world. A group called "World's Air Pollution: Real-time Air Quality Index" (waqi.info) receives data from over 10,000 remote sensors all over the world (the oceans and North America north of southern Canada are the largest blank spaces). They measure a number of the key air particulates of concern and rate them in six categories:

<50 PPM	Green	Good
50-100 PPM	Yellow	OK for most; moderate health concern for a few
101-150 PPM	Tan	Unhealthy for some; could affect healthy people also
151-200 PPM	Pink	Unhealthy for all
201-250 PPM	Blue	Health Effects for All
251-300 PPM	Purple	Hazardous, don't go outside

On their daily display overlaying the world map, it usually looks like a sea of mostly yellow and tan, with areas of green but also large pockets of pink, blue, and some purple. The U.S. is mostly yellow, with some fair-sized green areas and a few small blue areas. New Zealand, Canada, and central Australia are about the only places that are mostly green. India and China are mostly tan, pink, and blue (plus several purple).

A recent report from the Teton National Forest, using data collected over a 30-year period in the Wind River and Greater Yellowstone ecosystem, is chilling. Even in this supposedly pristine area, the level of nitrate in the atmosphere, which falls to the earth in the rain, is beginning to change the animal and plant biota on both water and land. The nitrates and other VOCs, of course, originate from faraway places as pollution, but when they fall into alpine lakes in particular, they have raised the nitrate level above the 'Healthy' level.

It has particularly diminished the diatoms in some lakes, with eventually far-reaching effects on fish and other biota, and also nearby lichens. Some alpine flowers, which appear to be extraordinarily sensitive to higher nitrate levels, are declining throughout the non-forest areas and being replaced by various weedy species or grasses, like cheatgrass. The ozone concentration, the main harmful ingredient in smog, is— surprisingly— also elevated in this high-altitude air. It is, in fact, above the level considered healthy for many types in the population: the old, young children, people with asthma or other allergic conditions, pregnancy, etc. The vaunted pure air of the supposedly pristine wilderness turns out Not So Pure. (We will look again at the ozone problem in a bit.)

B. Greenhouse and Other Gases

One of the most potent greenhouse gases is methane. Over the past eons of the earth, methane has sometimes been the major factor in causing the long-term global warming episodes (as well as some extinctions) in interglacial periods. Methane clathrate is the main storage form and is the single largest potential source of atmospheric methane gas. This 'frozen' compound is found mainly under all the oceans as well as peat, tundra, and the like. If this continues to thaw, releasing the methane gas, we are sunk.

Methane is now ominously increasing in concentration in the atmosphere. It is increasingly a major factor in greenhouse gas warming, along with other primary atmospheric pollutants, like CO_2, sulfur dioxide, and other industrial air pollutants. In 1860, the methane concentration was 722 parts per billion but is now 1866 ppb, the highest in at least 800,000 years. The main new methane sources are cows, rice paddies, bogs, and similar wetlands, permafrost, and ethane leaking from natural gas and other industrial facilities. It is a ticking time bomb. If the polar and ocean clathrate melts, as it is beginning to do, it will mark a tipping point, and we probably could not stop it in time to keep the worldwide temperature from catastrophically rising. (If it isn't too late already.)

Currently, much of the methane gas, nearly 50% of new methane, is produced by 'methanogenic' bacteria. This peculiar group of organisms, Archaebacteria, was the earliest type of life to appear on earth. Many of these bacteria have a particular niche, a common one being in the gut of cows. Thus, the 1.6 billion cows on the planet are actually now the single worst methane polluters (about 30% of total new methane).

An interesting, wry example of how science and technology might be used to correct the environmental damages that we are causing is developing in this arena. A scientific group from the Sunshine Coastal University in Australia discovered a putative 'fix' for the entire bovine methane production problem. They found that a single addition to the diet of cows would almost completely inhibit or kill the methanogenic bacteria and, therefore, reduce their host's methane gas production to near zero. They found that adding 2 percent of a pink seaweed called Asparagopsis to their feed would accomplish this feat.

So now we have a conundrum. Do we go all in on collecting large amounts of this seaweed and distributing it all around the globe so that methane from the cow burps and flatulence is all but eliminated and no longer contributes to increased atmospheric methane? Scientifically, it would probably work. But how do you develop the energy-intensive infrastructure to gather megatons of a relatively scarce seaweed, then distribute it to individual farmers for them to add it to their feed? What about the additional fiscal costs, to say nothing of the new intrusion on a particular ecosystem, and the fossil fuels expended at every turn? Also, it would probably create a physiological problem inside the cow, with likely negative effects on their overall health.

The 'correct' answer is obviously not clear, but here is the counterargument: going all-in on the technical side to eliminate a problem in 1.6 billion cows is a gigantic undertaking and could lead to some beneficial results. This, of course, is the general kind of technical "hard path" approach that has been taken in virtually every other similar choice situation for the past several hundred years.

However, why will we continue to maintain an earthly herd of 1.6 billion cows? Human livestock actually makes up the largest single part of all animal biomass on earth. In addition, George Weurthner (2012) says, "Livestock is the largest single cause of deforestation and sedimentation of coastlines." Keeping this number of livestock, therefore, would seem to mean chasing the same ever-expanding tail again. Impossible.

In any event, we (the U.S.) are already producing way more than enough meat to feed the U.S. and a sizable portion of the world besides. Furthermore, it is almost universally agreed that increasing per capita consumption of meat could be bad for both the environment and possibly for the person. We should be looking at means and methods to lower meat production, not devising 'fancy' techniques that will allow us to continue to keep expanding the meat supply. The World Resources Institute recently said that if Americans replaced one-third of their current beef consumption with some combination of grain, cereals, vegetables, etc., it would free up an area bigger than California.

If we would instead lower the number of people, we would need fewer and fewer cows making methane. It would seem, in fact, that trying to maintain a herd of anywhere near 1.6 billion cows is a foolhardy concept from the old school. But possibly a combination of both approaches would be the best initial approach. Go ahead and try the seaweed remedy on a limited basis, but while you are at it, vigorously pursue the most important part. That is, to lower meat production worldwide. Lowering the input of methane gas into the atmosphere, no matter the source is always a good thing. This example illustrates that, as with all of our big environmental and social problems, we cannot rely solely on our own ingenuity or technology to science our way out of it.

As noted above, there is also another recent alarming aspect of the methane problem. The arctic permafrost (containing huge stores of methane clathrate) is now melting unexpectedly rapidly and is already releasing considerable methane into the atmosphere. The warming effect of this 'new source' of methane will be very bad and will soon create a tipping point and uncontrollable temperature rise. Trexler points out that "the ultimate result of the predicted temperature rise would make it highly unlikely that human life (and much of the other plant and animal life too) could survive."

("Is there no end to this kind of stuff?" you may be thinking. So are we. But only we can end it. That is what this book is for.)

Nitrous oxide is another potent greenhouse gas (300 times as potent as CO_2). Another nasty effect is the depletion of ozone. The nitrous oxide concentration, too, is rapidly rising. In fact, the curve of nitrous oxide is going virtually straight up (like the population curve). Currently, it is the highest it has ever been (at least in the past million years) and stands at 335 ppb. Its source nowadays is mostly from industrial processes such as fertilizer evaporation, solid wastes, and fossil fuel combustion.

Ozone (O_3) is another highly significant gas. In the air, it has a 'funny' history. It is actually, to some extent, a 'good' gas. It forms a layer in the upper atmosphere where its main effect is to block excess UV radiation. The story of ozone is a mixture of good news-bad news. It is mostly a huge success story that actually saved the planet from horrendous catastrophes. In the 1970s, when the ozone hole was first reported, the world (amazingly) effectively responded. Scientific discovery, citizen activity, government action, and technical innovation finally came together for the Montreal Protocol in 1987. The principal fix was to eliminate the use of fluoro-hydrocarbons and similar chemicals, mostly in aerosols that were primarily responsible for depleting ozone, which created an 'ozone hole' in the upper atmosphere. The ozone hole then began to start shrinking and has become relatively stable.

This achievement ranks as the first, and maybe only, time that worldwide agreement and effective action were taken to tackle a potentially unprecedented global disaster. Let's hope it is not the last. The continuing good news is that this is largely working. The bad news, though, is that the fluctuations in the ozone hole are still worrisome. But as long as the world accepts the fact that this problem needs to be continuously and effectively addressed, catastrophic effects will be averted. However, the continuing success is not at all assured. If it should again start expanding to any extent, that would have devastating effects on human and animal health and ecosystems almost immediately.

Elevated and rising concentrations of CO_2 are obviously currently the most concerning aspect of human-caused environmental damage. CO_2 emissions rose to 36 million tons in 2018, creating the current record air concentration of 417 ppm and rising. This is the highest in the past several million years. We won't go into the causes and effects of CO_2 pollution here, as this is well known (as well as willfully and irritatingly widely ignored). A recent UK Population Trust Report stated that the social cost of CO_2 is $85 per tonne. This cost, however, is not actually being paid, so it is not noticed by most people. Suffice it to say, hopefully, more active measures will be taken to solve the world's most famous problem. The "Velvet Touch" will have to be a part of the solution.

Again, we say: 'While climate change presents one of the clear and present dangers, nevertheless, trying gigantic technological, scientific, and engineering workarounds without reducing population will be a fool's errand.' Fewer humans would put fewer and fewer CO_2

molecules in the air, and the ecosystems will, over time, take care of getting the excess out.

Besides CO_2 and the other gases, there are other major pollutants that we are mostly responsible for putting into the atmosphere. Things like particulates, sulfur, mercury and other metals, dust, and other aerosols. A little-appreciated fact about trees, most especially, but actually, all larger plants, is that they also incidentally filter out many of the pollutants that we have put into the air. Not enough, obviously, to prevent most of our current health problems with such things as mercury, lead, VOCs, fine particulates, and the like, but 'they try.'

As we should all know by now, the fallout effects of most types of air pollution, along with all the other environmental degradations, have been getting worse. The effects of these on climate health and ecosystems are becoming quite well known and, to some degree, being attacked by a host of groups, countries, and individuals. However, it is, by and large, a losing battle. The Secretary General of the UN recently issued a dire warning to the Assembly: "The world is moving in the wrong direction — and is facing a pivotal moment when continuing business as usual could lead to a breakdown of global order and a future of perpetual crisis."

Chapter 6
Land, Food and Agriculture

Is there a problem with food? Well, maybe. Some say we have enough available to feed the world, now and also into the future. The problem, others say, is that there just isn't enough food available, and production needs to be stepped up. In fact, many say there is a crisis now, and there are looming crises with hunger and the ecology or war and refugee problems of worldwide scope. We will have a look at the actual data of food production—— and also the huge damage that is done in our quest for this food.

As the human population grew to a certain point (25,000,000? 50,000,000?), a new phenomenon happened. Suddenly, the impact of a single species dramatically expanded. One species began demanding the lion's share of their particular ecosystem. Not only that, they expropriated more and more food, space, material, and water from their supporting biota, creating disruption and lasting damage (like deserts, etc.) over large expanses. Then they began spreading into neighboring and even far-flung ecosystems, like an invasive species. Now, we seem to want (actually, we already have) it all: all the time! But now we are finally beginning to feel the awful effects of our food hunt in terms of problems with water, soil, air, etc.

Of the 15.77 billion "real" habitable acres, over 50% are given over for man's more-or-less direct, heavy, and continuous use. Agriculture occupies 60% of this 7.9-billion-acre area (about 5.9 billion acres). More than 50% of these acres are used for livestock's direct utilization, which leaves about 2.6 billion acres for farm and cropland. It is interesting, too, that human livestock makes up about 60% of the biomass of all mammals of the world. Humans themselves account for 30% of that biomass, while all of the other wild mammals account for only 4 percent. This illustrates the extreme inequality of man's appropriation of space and resources. There is essentially no more space to be taken as-is for any major extension of agriculture (or anything else).

Looking for the moment at just the U.S.'s contiguous lower 48 alone, we see that it comprises about 1.93 billion acres of dry land. Wetlands, swamps, etc., occupy about 215 million acres, and the four major deserts comprise about 296 million acres. About 1/3 of the total land (a whopping 650 million acres) is used for livestock and grazing and pasture alone. Forests and other wooded areas occupy about 850 million acres. Other agricultural land uses, mostly cropland, are over 500 million acres. In the U.S., 408 million acres are currently labelled cropland. In fact, in 2018, only 308 million acres were harvested for the 20 main grain and other food items, like potatoes, etc. (Erle & Ramankuthu, 2018). Thus, we see that almost 1.2 billion acres of the total 1.9 is used, one way or another, for agriculture.

The ecologist George Wuerthner (2002) calculated a finer breakdown of human, non-agricultural land usage: Table 6.1 below.

Table 6.1. Type of U.S. Land Use and Acreage

Type of Use Acres (in millions)	
State & Nat'l Parks	44
Wilderness	64
Farmsteads	8
Rural roads	3
Airports	3
Railroads	3
Golf courses	2
Development & Rural Residential	239
"Defense"	25

Thus, a good chunk of land has been appropriated into urban, exurban, or other developed areas. And this does not really grasp the insidious encroachment of people and their things into even the remote places. Sprawl of all varieties, from summer castles in the wildlands and forests to resorts, subdivisions, roads, fences, and all the rest, is everywhere. Sprawl acres in the U.S. increase by about two million acres every year. It is probably more accurate to assert that most of the country is chopped up and 'bothered' enough by man to drastically affect all local ecosystems. Land has been exquisitely parceled out, fenced in, and roaded by, which are some of the little tools man uses to disrupt the intricate business of Nature as he single-mindedly pursues his own narrow economic interests.

A. Agriculture Footprint & Damages

The huge impacts of normal, large, monoculture agriculture in the U.S. is still largely unappreciated. However, it takes up to 80% of all the available fresh water, plus about 70% of the habitable land. Thus, its impact on the environment from virtually all standpoints, from wildlife, soil erosion and poisoning, traffic, and all the rest, is very great—overwhelming, actually. An essential problem, too, with current agriculture is partly inherent in the methods that farmers use to till the soil and raise their crops. Even worse is the scale of the enterprise. Wuerthner (2012) states that Big Agriculture is, in fact, by virtue of its enormous footprint, the leading driver of species decline and biodiversity in the US. He says: "Without a commitment to population reduction and a scaling back of industrial agriculture, the planet's ecological integrity remains in serious jeopardy."

Perhaps most damaging to the environment by agriculture are the vast swaths of land harrowed and harried to fit the needs of monocultural megafarms. By this devotion to maximum production of often nonnative, single-strain plants, the original ecosystem is essentially torn to shreds. The usual practices of "big ag" also require, in many places, perhaps the majority, of huge water, chemical, energy, and other inputs. A 2019 Report by the UN FAO (Belanger & Pilling, 2019) contained stern warnings about the decline in genetic biodiversity due to a large extent to current practices in agriculture. A 2021 FAO Report stated that "34% of agricultural land worldwide is now seriously degraded."

The soil also just happens to be one of the largest sites for sequestration of carbon. The poisoning of the soil by pesticides, fertilizers, salinity, herbicides, and the like is also taking a huge toll on the "little people" of the soil. These are the earthworms, decomposers, ants, and other insects, fungi, and the animals, including birds, that ecosystems normally depend on in no small measure. To the numberless little people in the soil, the bacteria and earthworms, etc., humankind, indeed all the biosphere, owes the biggest debt. They are the ones who 'toil in the soil' to essentially make the tilth (the soul of the soil). They have been hit hard in the past century.

The worst part of all this at the moment may be that the frenzy to feed the world continues to cause people to rush to expand food-raising acres worldwide. Most of the available land in the U.S. is already spoken for, but in the tropics especially, the race to cut down the tropical forests (the richest ecosystem in species and biodiversity on the planet) is in full swing. Besides logging, even the local, mostly subsistence farmers are busy clearcutting the trees to make room for just a few more acres to grow their crops —— their only means of making a living.

Thus, just the math makes it clear that agriculture is now the main keeper of the land. That fact, given the modern practices of industrial-scale farms in the industrialized world, mostly monocultural at that, makes obvious that the original ecosystems have been, at best, chopped up into discontinuous pieces. Crops and pasture, fragmented and busily used, are a far cry from a total functioning ecosystem that previously was there. This alone is antagonistic to one of the main precepts of ecosystem wholeness. Tom Philpott (2020) says, "Big agriculture is consuming the very ecological foundations that support ag itself."

Perhaps the worst of our ag problems, eventually, may be soil erosion. Soil erosion is the 'silent killer.' It is the one single thing that may wind up causing the worst of the coming dance between keeping even a few billions fed and unimaginable ecosystem dysfunction and loss. Soil is being lost at the rate of over one-quarter inch per year in the intensive- extensive agriculture areas of the U.S. (It isn't much different in most other parts of the world either, [Verso, 2015]) Incredible as it seems, on average, according to the USDA Resources Conservation Service, the U.S. has lost 6.8 inches of topsoil in the past 200 years. The best topsoils in Iowa are only about 40-45 cm deep and rebuilds, under favorable conditions, only at about 0.5 tons per acre per year, according to Dr. Richard Cruse at Iowa State University (cf. Philpott, 2020). Lawton (2017) and Lang (2006) also finds average soil loss amounts to nearly 6 tons of 'dirt' per acre per year. A leading soil expert, Pimentel (2006), and others predict that most of the topsoil in a great many places will be essentially lost in another 60 years at this rate. Then what?

Similar conditions, or even worse, are occurring all over the world. In fact, the loss in a great many areas of most countries is ten times the replacement rate of soil, even under the best scenarios (Montgomery, 2007). Verso compiled a table showing the amount of area badly affected by erosion in all countries. The world average loss was 35 million square km, while in the U.S., 3.1 million square km was lost. Thirty-six billion tons of soil was estimated as the annual loss of soil over the globe. David Pimentel considers soil erosion the number two problem in the world, close behind overpopulation. "Soil, the foundation of civilization, is crumbling."

It is, of course, not fair to pin all the fault on farmers. The U.S. Dept Agriculture's official policy for decades has been "farm from fencerow to fencerow." The sheer number of 50 acres here, 500 there, plowed up for big-time production farming leads to the present condition where two-thirds of available land is now nowhere near a functioning ecosystem. There are simply too many 100 or 1000 acre fields of a kind of half-ecosystem to maintain the populations of plants and animals that used to occupy those huge chunks of habitat and perform the usual services of a functional ecosystem.

"But," you say, "we need that production for food. We need to trade with and feed the world." Actually, we probably don't. At least on that scale, or for very long, if we would only work to solve the pressing problems in the developing world to break the back of their crippling food and population problems. One should ask, should it be the goal, then, to focus huge efforts to continue to try to feed the world no matter what the cost to earth as well as our own lives in the future?

Actually, we really are not bending all efforts to help feed the world anyway, as is often touted. For one thing, about 80% of the annual U.S. corn crop goes to making liquid biofuels and other non-food items. The same is true of soybean crops, with about 90% going to biodiesel and other areas of use. In fact, as Robert Bryce argues powerfully in his book "Gusher of Lies", the whole idea of biodiesel and bioethanol for fuel is now best seen as basically an ill-considered scam. Calculations of total energy and CO_2 balance by ecologists also have shown no net benefit, as much more fossil fuel energy is expended in their production than the final product yields. They are surely not going to be of much worth in even the short term and certainly not useful as a major energy source in the future either. Thus, if we really were laser-focused on feeding the world, we would, one would think, not sabotage the efforts by such dubious and soon-to-be-outmoded (stupid, actually) activities. The current governmental programs would seem to be inimical to a principled (sane, even) ag and energy policy now and especially for the very near future.

B. Big Monoculture Agriculture

Let's, then, take a look at the basic agriculture production numbers. Over half of the world's food supply comes from grains. The top 15 grain and cereal crops around the world are all of a very restricted genetic base. The most common and most important food crops that are used around the world and the total production is listed in the following table (Table 6.2).

Table 6.2 In order, are the world's top crops and their annual tonnages.

Crop Production (millions of tons)	
Corn	712
Rice	630
Wheat	625
White potato	321
Soybeans	215
Cassava	208
Barley	138
Sweet potatoes	123
Sorghum	60
Millet	31
Oats	24
Dry bean	23
Rye	15

The total world tonnage of these grain and cereal crops stands at around 2,857,000,000 metric tons or about 6.5 trillion pounds.

Of all these, four crops are the main ones, e.g., wheat, corn, potatoes, and rice. Notably, most of these are monocultural and produced to a considerable degree from 'Big Ag'. The origins of Big Ag, or industrial-scale agro-business, is very recent and arose innocently enough. Feeding the world and becoming the world's breadbasket were ubiquitous slogans and had, in fact, been the dominant mindset. Farming fence row to fence row was essentially the felt and also official mandate since the '40s, and this fit right in with the zest for fast expansion and growth in all things economic and material. Higher productivity and progress were the main drivers of not only agriculture but all aspects of the age, e.g., the (increasingly dystopic) "Human Game" of McKibben (2019). Now, we have a real problem because of all this.

The current ag practices stem to a great extent from the Green Revolution in the 1940s, '50s, and '60s. Technology was applied on a global scale to improve not only the practices of farming but specially to develop high-yielding, disease-resistant strains of the main food crop species. The Green Revolution initially arose, in considerable part, from the approach and work of one plant scientist, Norman Borlaug (the "Wizard" of Mann). Beginning in the late 1940s, he worked on a small, very poor, experimental field in Mexico and decided to try to breed a wheat strain that would grow well there and, especially, be resistant to the deadly wheat stem rust. Doggedly, under unpromising and near desperate conditions, he kept working on an increasingly ambitious goal of breeding miracle wheat. A short strain (prevents 'lodging'), resistant to wheat stem rust, with larger, more nutritious heads and much higher yield than all previous. By the early 1960s, he began to have success in field trials in various places, including Mexico, India, and Pakistan.

His miracle strain of wheat became a 'thing', and he was awarded the Nobel Prize in 1970. Unheard-of yields were obtained in several fields all over the world, and this miracle wheat essentially became the standard. Initially, though, in Asia particularly, farmers found that their yields, while higher than before, were not the astounding yields advertised elsewhere. However, Borlaug had insisted from the start that to obtain these high yields, one had to also do other things. Like vigorous field preparation, generous nitrogen (fertilizer), keep weeds out, and so on. And he was right.

This is the "hard path" that agriculture has taken. Soon, other miracle crops, like maize and rice, were bred or engineered and adopted worldwide. New strains of higher-producing rice were pursued, particularly under the aegis of the IRRI (Institute of Rice Research Institute), funded by the Rockefeller and Ford Foundation. This effort has resulted in the breeding of the current crop of quite good, high-yielding rice.

Similar, parallel "hard path"' approaches to filling the perceived urgency of expanding agricultural production to keep up with growing humanity have been proceeding apace all over the world. Genetic engineering has been done to increase desirable characteristics for many species. A very impressive example has been 'Round-up Ready' corn, alfalfa, and others. This was a brilliant idea. Noble task! It is a wonderful boon to not have to worry about weeds taking over a newly planted field, and it increases production marvelously. It makes the work of farmers less and the profits more. What could be better! A win-win on a grand scale.

However, the herbicide Roundup (or any herbicide) is not benign. It is increasingly under fire from medical as well as ecological fronts. It is a suspected carcinogen, for one thing. It is now known to be toxic to birds, insects, and other microbiota of the soil. Its long-term effect on the environment is increasingly under study and suspicion.

Further, the new 'miracle' crops usually need constant generous applications of pesticides, herbicides, and fertilizer to keep the enterprise going. All these not only do not add anything directly to the soil or ecosystem but also require a great deal of extra material, especially energy from fossil fuels. (Interestingly, chemical fertilization plants produce more nitrogen than is fixed by all terrestrial plants. Also, perhaps not surprisingly, only four gigantic multinational corporations produce, for example, most of the total fertilizer produced in the whole world.)

In addition to the genetic and technical improvements engendered by the Green Revolution, new and larger methods and equipment for farming were necessary to take full advantage of the improved plants, mainly with the laser-like goal of maximizing production. Massive tractors and implements on a large scale, with new, faster-growing, higher-yielding crop strains, which now required more herbicides, pesticides, and especially fertilizer. This virtually drove out small farmers, and Big Ag now rules in many places around the globe.

From the beginning, seed companies started making better, i.e., higher-producing hybrids. Better seeds meant more farmers using them. A byproduct of that was that larger and larger seed companies controlled all aspects of ownership and distribution. So now, virtually all 20-some of the major seed crop types are of just a handful of varieties, used by virtually all farmers and manufactured by fewer and fewer corporations. All of this has drastically reduced biodiversity and resulted in the extinction of thousands of fruit, vegetable, and cereal crop species all over the world.

Farm machinery, too, became bigger and bigger (colossal even, one might say), made and sold by only a few companies. Fertilizers and pesticides, herbicides, chemicals, etc., also became essential on a gigantic scale. These, too, became consolidated by a few large companies. Thus, this industrial monoculture agriculture grew up and expanded until we get today's mega Agribusiness and monoculture. This is the way our current farming practices came into being. The virtual squeezing out of small farms is indicated by the fact that over 60% of crop harvests in the U.S. are from corporate or very large farms, even though they only comprise about 4 percent of the U.S. 2.02 million farms. And, of course, the whole Green Revolution enterprise was subsidized by fossil fuels.

Basically, the principal process involved in all the above developments amounted to the wholesale industrialization of the entire food production system, spurred by the growth of the population. This approach has spread to factory fishing, too, with sophisticated trawlers and battery farming for pigs, poultry, cows, etc. All this was seen by us, but not really, just more or less unconsciously (ala Goleman's and McKibben's observations). It didn't really register. These were the things, of course, that our generations and those before us thought were only the smart and obvious ways for us humans to operate—'Grow and expand production and profits, endlessly'.

This "hard-path" agricultural effort was wildly successful to this point, obviously, and has produced what modern agriculture in developed countries consists of today. Half of the wheat grown now is comprised of only a few new miracle strains. Same with corn. Seventy-five percent of all global food production comes from only about two dozen highly inbred crops. And these are the ones selected for (indeed led to) the current monocultures.

The loss of genetic diversity in all these staples of the human food chain is, however, a major cause for alarm for the future.

Overall, looking back 15 thousand years or so, we see that agriculture once changed everything; then Civilizations and History changed everything again; then the Industrial Revolutions, then the high-tech Revolutions changed it all again. Humans had to adapt and change a lot, and we did, actually, culturally evolve— and very rapidly, too. We could optimistically actually detect overall an improved social human nature (a more domesticated Homo sapiens?). We apparently have fewer wars or famine and distress in so many places (cf. Pinker, 2011, 2019), so we can take some little credit for that. We should also, however, acknowledge the damage we have inflicted on the earth unwittingly along the way. All these centuries, it appears that everybody had been peering straight down into the furrow we were plowing. The furrow now is a deep and wide gash in the earth, rapidly becoming unbridgeable and ultimately lethal. Everybody has basically 'forgotten' to add the environmental costs to the cost of food (and everything else).

C. How Much Food Is There?

We'll start with the situation in the U.S. Currently, there is available about 400 million acres of cropland, which produced 1.9 billion tons (3.8 trillion pounds) of the main crops in 2017. These gigantic crops were harvested that year from only about 308 million of those total cropland acres. These are the totals of all the 15 important ag crops, such as all grains, corn, soy, potatoes, rice, etc. These are essentially the basic foods to feed the country and also to use for trade. (Along with meat, nuts, fruit, and vegetables, these comprise essentially all domestic foods.)

But the human food that is actually available for people from all 3.8 trillion pounds of these crops is only a small fraction of this raw total production number. That raw tonnage of cereals, grains, potatoes, etc. (3.8 trillion pounds) does not constitute the 'real' amount of this food group that is actually available for food and human consumption. First of all, this harvested plant material also has to feed all of the country's livestock. Livestock in America consumes, in fact, over half of the U.S. grain crop.

In addition, we have not factored in the losses in these crop production numbers, nor the fact that a good deal of harvested crops goes to various non-food uses, like high-fructose syrup, biofuels, and many other types of uses. For example, about 40% of corn currently is used for bioethanol production alone. Only about 10 percent or less of the total corn crop actually eventually goes directly into human food. Similarly, almost all soybeans go for animal feed, oils, and other niche markets. Only some of these crops go directly into the food market, while most go for biodiesel and various other uses.

So, factoring in all these (major!) deductions, we find that the actual total grain and cereal production that goes for food is, at most, only about 950 billion pounds per year. That is still about 3000 pounds per year, or about 9 pounds per day, for every American. (The average daily intake of Americans is about 3-5 pounds.)

Next, then, to get an estimate of the total agricultural production, we can add in the total of meat, nuts, fruits, and vegetables produced commercially in the U.S. Meat production totaled about 103 pounds, and nuts, fruits, and vegetables amounted to a total of about 110 billion pounds. The U.S. total production of the major food types, therefore, apparently adds up to be 1,163,000,000,000 pounds or a little over one trillion pounds. However, there really was not that much "real human food" in this total. For one thing, we have not factored in all the losses and other uses, like dog and cat food, etc., from the total. The actual number of pounds going into all these losses and types of uses is pretty much unknown, so we will just arbitrarily use these raw numbers as the totals. (We have also not included auxiliary sources, such as fishing, hunting, gardens, etc.)

Let us recapitulate:

As we saw, the current US agricultural production of actual available food from all sources totaled 1.16 trillion pounds. Obviously, that is far more than enough to feed ourselves. In fact, that is equivalent to 3400 pounds for every man, woman, and child in the U.S. (Again, the 'normal' food intake per adult per day to maintain health and weight and grow is 3-5 pounds per day.)

The actual (or estimated) amount of food exported was nearly half a trillion pounds for worldwide distribution. On its face, then, American exports would supply only 65 pounds per year for each person around the world, i.e., not very much.

Thus, we note that American Agriculture today comfortably supplies the U.S. population with plenty of food, plus much more to spare. It is way more than needed to feed the country and to theoretically also provide a meaningful amount to help feed the world. In actual practice, of course, a great deal of it does go into the international trade vortex. This labyrinth is probably understood by someone somewhere, but it seems to be kind of messy at best. Certainly, the distribution of food is spotty and unequal, to put it mildly. Even so, American surpluses are far short of being enough for the world, too.

So, we have answered two important questions:

1. Can America readily feed itself with present agriculture?

The answer obviously is 'Yes, certainly, with a goodly surplus of most food commodities to boot.' (But we also have seen that this does come at a substantial, and actually nonsustainable, cost.)

2. Can America feed the world?

The short answer is "No, not even close." Certainly not without major enlargement of the ag enterprise. But still, we are the major breadbasket and a huge factor in helping to combat malnutrition and starvation around the world. We are, in fact, quite nicely producing surpluses for our own country for now and the immediate future. And we make a little dent in the food deficit that has been facing the earth for the past century.

3. Can the World Feed Itself?

How much food is there worldwide? We estimated the total amount of food available in the U.S., but to answer that for the world is a harder task. Various estimates by FAO (FAO.org "World Food & Agriculture Status Pocketbook, 2018) and other groups (USDA, Agriculture Outlook.org.) have been from 6.9 to 11.2 trillion pounds. Other estimates given by FAO.Org/SaveFood Global Initiatives on Food Loss & Waste report for 2016 were around 8.1 billion tons (or 16 trillion pounds).

Let's take a middle-range estimate (11.2 trillion lbs). That equates to around 4 pounds of food per day, enough to theoretically feed everyone on Earth and then some. (Americans consume a ton of food per year—the actual figure in 2018 was 1996 pounds.)

So back to the essential question: is there enough food produced somewhere to feed everybody now? All in all, the numbers would seem to give a little comfort to those who say we're at least close to meeting the goal, so if we made efforts to increase production, maybe we could eliminate hunger entirely. That is if distribution went smoothly and equitably, waste was kept minimal, and weather and other conditions remained reasonably favorable (and if we also continue to ignore the continual population growth (75 million per year) and the growing ecological calamities on our doorstep).

The World Bank, however, calculates that food production worldwide will need to rise 70% to feed the predicted 9 billion people by 2050. (Interestingly, they also recommended not expanding "big ag" but says small, ecologically sound farming is the best route forward.) Another 2016 UN Report flatly stated that "worldwide food production needs to double by 2050." Most people more or less blithely assume that we definitely need substantially more, ad infinitum! (From the old reliable Cornucopian magic storehouse, perhaps?)

However, it is an inescapable, stark fact that at least 795 million people today and for some time to come will be hungry. (That is actually the largest number of inadequately fed people ever, even though it is also true that the percentage is not the highest.) Just to ensure that those people are supplied with 3-4 pounds of food daily would take a huge, concerted effort. By then, however, there would be 9 or 10 billion mouths to feed. All gains in productivity or distribution or waste reduction would again have been cancelled out, and there would be, again, not quite enough. Any putative gains would not be enough to reduce the deficit.

So, what these data really suggest is that, no, the world cannot feed itself very well or for very long. There is also no reasonable suggested pathway that would get us to that goal at present population trends (notwithstanding that the population growth rate is beginning to slow in many places). Even if we were to, say, more than double the food production, what would be the cost in terms of the already over-stressed biosphere, hydrosphere, and lithosphere? The answer is an environmental disaster in an even shorter order. Thus, this oft-suggested pathway is seen to be a fool's game, if not an outright cornucopian scam.

There's one more fact that needs emphasizing again. During the past decade since 2010, the total world food grain production has remained right around 2.2 billion metric tons (4.8 trillion pounds). In fact, while it is mostly steady, the trend is slightly downward. All this while the earth gained about a billion people.

In sum, as we have seen, and as many authors have pointed out, there seems to be almost enough organic material (food) being produced around the world today to theoretically feed everybody, even though the awful actuality is that we are dancing on the edge and aren't feeding everybody. But it is true that we could, if major efforts were made to quickly and efficiently improve the logistical problems all over the world, especially in the net importer countries of the basic foodstuffs. This, however, would be of benefit only for a short period (until more population growth makes it again insufficient) —— and the environmental toll comes rolling in.

We have to, of course, make every effort to feed our people and others in need around the world now and into the future. But how to do this without breaking Nature's bank is an existential puzzle. Instead of following the reasoning used in the past, though, we should ask, "Why not help all countries grow most of their own food and not have them depend on us? Why not begin to reduce food demand, i.e., population, where the demand is highest?" The whole food enterprise these days is driven almost entirely by economic, and not ecological or rational factors, or even with any serious considerations to physical limits. This is the mindset that must change.

That other—enormous— facet of food production is simply ignored, e.g., the negative cost that agriculture (and other frenetic growth sectors) already charges in terms of environmental degradation. Expanding an already too-environmentally costly enterprise and eventually still falling short of feeding everybody in the world, is that a really good idea? As documented earlier, loss of soil and water and ancillary pollution of all sorts are already taking an unacceptable toll. A toll that is still largely unseen and rather poorly known in detail but clearly seen by all who look with clear eyes. Increasing agriculture to any great extent would seem to be an absolute fool's game. Even if it were possible that with some modest improvements and expansion of agricultural operations around the world, we might (at least theoretically) continue to feed the equivalent of our present or slightly larger populations for a while, the ongoing damage would be grossly unsustainable and just make things worse. (Actually, the U.S., in particular, stands to run out of phosphorus reserves, for example. This, of course, would be fatal to Big Ag in a very short time.)

So again, we have to say (loud and clear) that we really do not have a sane, satisfactory pathway to feed ourselves and the world, too, for the foreseeable future. Certainly not in a sustainable way. So, what to do? More of the same? Try to drastically increase food production (and all the other growth areas too)? Same old? Same old? Madness! Why would governments and people willfully continue this Alice in Wonderland treadmill?

How about, then, we feed a smaller world? Say, a world with 1 billion people? What would agriculture look like in carrying out this new task? We will look later at some possible beneficial changes in agriculture that would seem to fit quite well into that new, hundred-year scheme. Of course, if population reduction actually began fairly soon, it would take the rest of this century and more to complete meaningful reductions in the scope of agriculture. This, in fact, is the reason for this book —— to ensure that the human population gets down to its comfortable carrying capacity, i.e., plenty of food for all.

The big payoff for all this, if and when we would reach such a goal, will be the land and ecosystems that would be restored. In addition, we'd have, at last, a sure and safe supply of good food for all the world's citizens. This would mark a milestone in history for all time. Truly, this is a noble goal to shoot for. That day is a long way off, of course, say a century or two? (Again, however, it should begin to be seen as an obvious fact that if we don't begin this kind of program, then at some point quite soon, any kind of sane equilibrium between people and the earth will have become exceedingly difficult, if not impossible.)

Paul Roberts in "The End Of Food" sagaciously capsulized the whole problem from top to bottom. He traced 'farming', i.e., supplying the food to feed the extant population, from the first tribes to now. Hunter-gatherers needed many square miles to feed their tribe. When agriculture began just 11-12 thousand years ago, food could be produced to feed more people on less land. Hence, in the path from villages to cities to societies to civilizations, each stage needed more and soon even more farming. More farming for more people needed more land, and on and on in an ever-increasing spiral. Along with 'more' farms, the need for 'better farming' then came along, needing less land per person but more land for more people. Thus, Roberts said, "Farming is one of the biggest polluters, and in view of its toll on waters, soil, and air, it calls for a rethinking of agriculture and the future of the world's food supply."

Now, if truth be told, we've already run out of suitable land for farming for even our current population without paying a catastrophic price. Thus, we are witnessing today the beginning of the end of this process; the Malthusian crisis is about to finally get us. Gouging out more land or cutting up more of the tropical forest can no longer be tolerated to make "more farming". Even inventing 'better' farming in the future would wind up with diminishing returns.

We have reached Malthus's crisis:

"The power of populations is so superior to the power of the earth to produce subsistence for man, that premature death must in some shape or other visit the human race."

–Thomas Malthus

The immediate problem, though, is how to keep feeding not only ourselves but a goodly number of people elsewhere with the current practices of Big Ag. Whatever we come up with, however, we should understand this can only last on an emergency basis for only a short time. There will come a time soon (yesterday?) when even Big Ag can no longer fulfill the need. All the while, of course, the ecological and even human toll will become increasingly intolerable. This is the familiar pickle we are in.

Chapter 7
Energy (Im-) Balance

So, let's now take a stern look at the state of "Energy" in the world. Ultimately, all the energy we use comes from the sun (with a few exceptions for nuclear radiation, deep earth thermal vents, pools, chemicals, and the like). Only one species has learned to harness nuclear energy, while a few, mostly marine types of organisms, have "figured out" how to harness direct non-solar thermal energy.

Solar energy, of course, drives photosynthesis and, thus, essentially the entire biosphere. Energy from the sun is used for many other actions, too, such as driving the wind and storms, heating the planet, weathering rock, and so on. Most (88%) of it is radiated almost immediately back into space as heat, keeping the earth (normally) at its usual (at least for these past several thousand years anyway) 'constant' average of around 57 degrees F. (For the year 2022, the Earth's average was nearly 2 degrees higher, e.g. 58.9 F.) A variable amount, but actually only about 12 % of the sun's energy, is currently retained at the surface to carry out all these normal earthly activities and also for permanent use, such as by plants and indirectly by all. Photosynthesis, then, essentially powers the whole of life's activities all over the earth all of the time. So, it is useful to keep in mind that all life depends on the sun and the sun harvesters (plants) exclusively for food and fiber.

Besides the energy we get directly from the sun, we presently still have at our disposal larger stores of fossilized and stored sunshine. Humanity discovered this many centuries ago, but only in spades for the last 300 years. Coal started the Industrial Revolution, and now oil and gas continue to feed it——at hyperspeed too. As is (hopefully) evident to all (a few?), these will eventually (actually quite soon) run out. Leaving us once again to have to rely on just our natural, renewable sources. This next time, however, we are looking not only at increasing demand but less and less stored energy and many other of Nature's necessary things. Something's got to give. Doing business as usual for the next centuries is a certain recipe for great suffering for man and nature both. Obviously, everything we use from some point, very soon, on will have to be from a sustainable source—and with much lower demand.

A. Overview of Energy

So, basically, we all live off the sun's energy. How much of that is there? Would we be able to sustainably get all we want? What are the collateral costs of our getting it?

The solar power (insolation) that hits the surface somewhere on earth, land, or sea all the time averages about 164 watts per square meter per day. Since the surface of the earth is 510 trillion square meters, that yields a maximum for all earthly insolation energy of about 83,000 terawatts (TW) per day. Over a full year, this equates to about $30,500 \times 10^{15}$ watt-hrs per year (or 30.5 Exawatthrs). That is the limit that could ever be (theoretically) harnessed. How much energy, then, is actually available to people, and how much of it is actually used? By whom and how?

The only original users of sunlight before animals and man came on the scene were plants and their kin. The amount of sunlight they use actually has a number—and a name. The amount of sunlight that is collected by all the earth's plants is called the Gross Primary Production (GPP). The 'people' who do that, of course, are the photosynthesizers, the algae and plants, and a few other odd microbes. It has been measured that all the plants in the world only use ~1.6 percent of the total insolation/year. This translates to the equivalent of about 488 Peta (10^{15}) watt-hours of solar power used directly in all photosynthesis per year. For comparison, man uses all around the world about 2.7 terawatts (TW) of electrical energy on an average hourly basis. This amounts to 23,600 Twhrs per year = 23.6 Petawhr. This isn't that much less than all of photosynthesis (about five percent, in fact).

The amount of solar energy that 'stays' in the photosynthesizers as 'protoplasm' is called Net Primary Production (NPP). The energy difference between GPP and NPP results mostly from the respiration, metabolism, etc., of the plants. NPP essentially equates ultimately to the total extant biomass of the planet. Haberl, Erb & Kaufman (2010) call the human fraction of the global NPP the HANPP (Human-Appropriated NPP). They found that humans now co-opt over 40% of the world's NPP. They say that this figure is a good indicator of human pressures on ecosystems. Using somewhat similar calculations, Ewing et al. (2000) estimated that this human ecological footprint equates to needing at least 1.4 more earths just to attain some sort of minimal equilibrium.

The actual amount of new plant material made with the sun's help over the whole mass of the land of the world comes to about 264 billion tons per year, or 528 trillion pounds (Vitousok, PM et al. 1997; Imhoff, M.L. et al. 2004). We may compare this total to what we estimate to be the total food production of the world, which is between 7 and 16 trillion pounds. Thus, we see that 'Nature' makes only about 40 times what we puny humans can do with our tractors and fertilizers. That seems strange ("scary") that all the plants of the rain forests and grasslands and woods and prairies amount to not that much more than man's own foodstuffs.

There is also a measure of the world's total amount of energy available and/or used by Man each year——the Total Basic Energy Supply (TBES). In 2019, the TBES was measured at 158 PWhr (=158 x 10^{15} Whr). The vast majority of that energy, however, was from fossil fuels. Obviously, that source will soon enough run out, and all of that source will soon have to come from some other (renewable) source.

Electricity is increasingly becoming the desired go-to form of energy for people, even though it currently accounts for only about a fourth of all energy used. Electricity is the most efficient, but its production obviously comes entirely from man's effort. (Ultimately, of course, that too comes directly from nature.) So, how much electricity is used worldwide? The nameplate capacity for electricity is listed as 6.14 TW. However, the average actual production was only about a third of that, or 2.7 TW (terawatts). The total world electricity consumed, therefore, amounted to 23,504 Twhr in 2018, as noted earlier. We should keep in mind, however, that at present, electricity accounts for only about one-third to one-fourth of the total energy usage. Most noteworthy, though, is the fact that almost three-fourths of all this electricity came from fossil fuels. Coal, oil and gas supplies 72% of that. All of the renewables combined produced 6 percent of the electricity.

It should be obvious that these fossil fuel uses need to be immediately cut back drastically. This, of course, is one of our urgent suggested tasks for our approach, which is to reduce demand by reducing the population.

B. Summary of U.S. Energy Use

The total U.S. energy usage in 2017 was about 101 Quadrillion BTUs (Quads). This includes all kinds of energy. This total of 101 Quads is equivalent in electrical terms to 29.5 x 10^12 kwhr = 29.5 Terakwatt hrs. Drilling down a little on the general picture of energy in the U.S., we see that of the total energy use in the U.S., the Big Three energy sources (coal, oil, gas) accounts for 76% of all energy. Renewables produced the rest.

The U.S. nameplate electrical capacity (i.e., the maximum amount of energy that could be generated at any one time) is around 1.2 TW. The total electrical use in 2017 was 4.18 trillion Kwh (= 4.18 Twhr), according to the EIA. Electricity, therefore, accounts directly for only about one-third of all energy used in the U.S. The electrical energy that is used per person in the U.S. per year is about 12,600 Kwhr. Every year, this demand increases by 0.4%.

In his latest book *How to Avoid A Climate Disaster* (2021), Bill Gates summarizes all these energy needs in a wonderfully clear way. He categorizes our energy use for five things: (1) "plugging in", i.e., electricity, (2) making things, (3) growing things,

(4) keeping warm and cool, and (5) "moving us around"'

Altogether, the energy cost, in 'CO$_2$ equivalents', amounts to 51 billion tons of carbon sent into the atmosphere each year. Making electricity "costs" 13.7 billion tons (27%). Making things like paper, glass, cement, plastic, steel, etc., costs 31%, or 15.8 billion tons. He calls the use of cement, steel, and plastic the 'big three', which accounts for 90 % of the energy for making things. Growing things amounts to about 9.7 billion tons of CO$_2$ equivalent and keeping us warm and cool another 7 %. Transportation accounts for 16% of the total 51 billion tons of CO$_2$. No matter how you slice it, however, our total use is unsustainable and, soon enough, fatal to much of nature, which includes us.

Table 7.2, below, is a summary of actual energy use in the U.S. (data from the American Geoscience Institute 2017 Report on U.S. Energy Use.)

Table 7.2

Source	(Quad & % of total)	Use For Electricity
Oil	36 Quad (37%) 1%	7.5 BB/yr;
Nat. gas	29 Quad (28%) 35%	38 trill. Cu. Ft.
Coal	16 Quad (17%) 93%	~700 million tons
Nuclear	9 Quad (9.5%) 98%	
All other Renewables	11 Quad (12%) 38%	

TOTAL Energy = 101 Quadrillion BTU's

Following is a list of renewable energy sources and their percentage contribution to the total U.S. electrical energy consumption (Table 7.3).

Table 7.3. U.S. Renewable Energy Sources

Renewables	(Percent of Total Energy)	TeraKilowatthr
Nuclear	(9%)	.807
Solar	(1.8%)	.070
Geothermal	(0.4 %)	.016
Biomass	(2%)	.071
Wind	(7.3%)	.300
Hydro	(6.1%)	.274

TOTAL 1.53 Terakilowatthr

Thus, we see that all of this (electrical) renewable energy amounted to only about 28% of all electricity (but only 20% of total energy use). This figure is about the same percentage as coal. Therefore, when coal is done (or we are 'done' with it), renewables will have to double to fully replace that 'void'. Of course, that is precisely the idea that we would like to see implemented rather soon anyway as part of our general population reduction proposal.

Presently, renewables are only increasing at a paltry rate of 0.4% per year.

The British Petroleum Company produces an annual Report, "Review of World Energy." Their 68th Edition, in 2019, stated repeatedly and unequivocally that the present trend of energy supply is "unsustainable"; period. Pretty strong stuff from the most oily baron in the world.

C. Energy in the Future

So, what are the prospects for getting our energy in the future? Where would it all come from? The EIA states that global energy demand rose 5 percent in 2018 and will continue to increase by another 29% by 2040. Their estimate for total world energy usage by 2050 came to about 27 terawatts (TW). To get current electrical output eventually to a (mostly) all-electric world, therefore, would mean increasing world electrical generation almost ten-fold. That would indeed be a tall order——fantasyland, actually, even if we should (stupidly) increase our use of fossil fuels.

Of course, most of all current energy (~70%) comes from fossil fuels. That means that the world needs to be preparing for not only greatly increasing our total energy output but also how to replace that with mostly renewable sources. Simply put, there is no possible, reasonable way to get to anywhere near the 27 TW estimated to be needed. There is no way we can maintain even our current usage.

What doesn't jump out from the statistics but constantly lurks behind all the scenes is that the whole enterprise is clearly unsustainable already. It is mostly done on the borrowed sunshine of fossil fuels and the stolen largess of various ecosystems. We all know (at least maybe deep down) about the damage that is being done by the digging and pawing all over the Earth to supply our current energy. The toll this is taking on the landscape as well as the biosphere, including global warming, etc., is already horrendous and unsustainable. Thus, considering that most of our energy is from a very finite storehouse of this fossil energy, it seems crazy to try to continue this kind of supply for 10 billion people indefinitely (even if that does seem to be the current, mindless plan).

There are only two possible remedies for our energy problem going forward: (1) make energy only from renewable sources, or (2) reduce the energy demand. (There is, of course, a third, urgently needed alternative——both of the above.) How is it, though, that so many people blithely assume that we can continue to keep increasing output and play the old familiar 'Human Game' for the next 30-100 years or so (or even better, forever!).

By the way, there is a lot of other stuff that we depend on that we are digging out of the ground unsustainably. "Enough platinum or titanium that we know of to last just another 50 years?" you say? "Hell, there's lot's more that we just haven't discovered yet," the smug and witless will say.

A (highly truncated) list of such non-renewables includes, for example:

Manganese	40? years left
Cobalt	10? " "
Copper	40-200 " "
Nickel	100? " "
Zinc	40? " "
Tin	100-300? " "
Gold	40 " "
Uranium	75-100 " "
Lead	35 " "
Platinum	40-50 " "
Iron ore	50 " "

Daniel Yergin ("Quest") has traced energy all around the world from day 1, from campfires to nukes, as he described man's historic reach for energy fuels. Coal and now gas and, especially oil, have occupied an outsized chunk of man's energy sources. They are the fuel sources that did the conquering of the planet. Along with these big three, he traces the history and near future of the other possible fuels, including nuclear. Here, in this discussion, however, we will ignore future nuclear expansion, as that is still so fraught with uncertainties and dangers of its own peculiar kinds. It seems highly unlikely that any form of nuclear source could be sustained over long periods of time.

Yergin surveys the future feasibility of all the other possible energy replacements, like hydrogen, etc. He traced the history of solar, wind, and biofuels essentially from their start in the 70s and 80s. "Gasohol" is currently the largest type of any of these possible candidates; only hydropower comes in a close second at present. Hydropower, however, is widely considered, even among environmentalists, as having already nearly reached its peak. Gasohol, i.e., bioethanol and biodiesel, however, is not really worthy of further consideration for reasons noted earlier. The other commonly available possible energy sources, i.e., solar, biomass, and wind combined, only account for 11 percent of total renewable electrical production currently.

Paul Roberts (2004), in "The End of Oil", describes the history of our increasingly insane tango with energy. He foretells the end, which: "apocalyptically or not, Will soon end… There are fewer and fewer reasons to believe that our overtaxed energy system won't have begun to collapse before then" [2050].

Yergin (among others) notes that major expansion of current energy sources, wind and solar, too, is fraught with difficulties. Windmills, especially, are large and complicated manufactured machines, after all, and also do not have a very long lifespan. Solar panels are also complicated manufactured items with substantial Life Cycle Analysis (LCA) costs. Yergin is very skeptical of any of these sources (including nuclear and, especially, liquid biofuels) being able to scale up to supply anywhere near the nation's or world's energy needs. Of course, the energy needs of the future that he used are always forecast to be rapidly increasing as well, which adds enormously to the challenges. (William Vollman [2018] gives us an interesting, sometimes hilarious, sometimes head-banging, two-volume, ankle-high survey of our last centuries' insane dance with producing and consuming our past and present energy.)

Yergin says the latest leader in the search for future big energy fuel sources is the so-called cellulosic biofuels from biomass (which is mostly comprised now of liquid biofuels, like ethanol). Importantly, he cites that the most promising new potential replacement energy source is synfuels. (We will look further at this source in the following sections.)

In any case, in looking at all the possible scenarios for future energy sources, two things seem obvious——(a) we will need to be weaned off fossil fuels, and (b) it will be increasingly difficult, if not impossible, to provide all this needed energy if present, or higher, levels of demand exists.

If, however, populations would start decreasing, then less and less energy would be needed going into the future. Even then, though, the future energy requirements per capita, i.e., the energy footprint, might actually need to increase a little since the new economy will probably be using energy in (hopefully) many new (environmentally friendly) ways. In the future, however, there will be less big machinery or little gadgets, and the physical footprints will become more "immaterial", with greater dependence on advanced technology, electronics, information, community actions, less frantic living etc. Most importantly, more emphasis on community and human life. Of course, we will be needing 100 percent renewable energy sources, too.

As noted earlier, renewables at present (including nuclear) comprise about 30% of the U.S. electrical generation, amounting to a total of about 1.53 Pwhr. We have also noted that the U.S. total electricity production is about 4.2 Pwh. But this makes up only about one-third of the total energy. So to replace our entire energy requirement with 'new' renewables, therefore, would require at least an additional 8.4 Pwhr. Could we, then, between biomass, solar, wind, wave, hydro, geothermal, hydrogen, and all the other options being considered (cf Roberts, 2004) actually achieve that goal? Wind and solar are presently being looked at as the major new future sources. Huge expansion of wind power, however, to a large extent, is considered a rather dicey and expensive proposition in terms of materials and footprint. Thus, what would happen if man today were to try to get his energy mostly from sunlight and biomass, along with nuclear, wind, hydrogen, and all the others?

1. We can start with solar.

Solar will surely be one of the important core future means of supplying our electricity. A recent NREL report (2016) concluded that the placement of solar panels on all the rooftops in the U.S. could provide 40% of the country's yearly electrical energy. Thus, if we were to get "all-renewables" to just the current 4.2 Pwhr of total U.S. electricity production, it would require solar to produce an extra 3.1 Pwhr per year.

Is that a doable goal? Suppose we assume a conservative average insolation factor and solar conversion efficiency of solar panels of about 180 Kwhr/per year per square meter. We can calculate how large the solar array would have to be. That turns out to be about 17 billion square meters or nearly 7,000 square miles. That is about 4.5 million acres; theoretically, that maybe could be doable; only, however, with a triple-double, gigamassive engineering project unmatched in human history.

Even if we did this, though, we still would have replaced only one-third of our current fossil fuel use. To supply virtually 100% of our energy needs would obviously require three times that amount of space or about 21,000 square miles of solar panels. Is that doable? Is that feasible? It would certainly involve an even larger, superhuman industrial effort and huge mineral resources and also produce some considerable 'scabs' over the earth's surface, involving much ecosystem damage. In practical and environmental terms, however, it would seem to be unthinkable actually.

Thus, these huge solar "hard-path" Big Energy approaches do not bring huge immediate smiles or comfort to the minds of many people outside the Industry. The bottom line, it would seem that taking the "hard path" for centralized "Big Solar" at an enormous, super-grid scale would not be feasible at the present time to replace the lion's share of future energy needs. It should be remembered, too, that solar installations would have to be essentially replaced two or three times a century.

Solar, however, at a smaller scale, is not only highly feasible but also highly desirable and is expected to become a much larger player in the whole energy picture. Also, if the population would significantly decrease, supplying energy by this (and other routes, too) becomes easier and completely doable. For example, in 1960, the U.S. population was about 159 million, and the electrical use that year was about .8 Terakilowatt hours (compared to 4.18 today). If we multiply that .8 Tkwhrs by four, that would give us a rough guess at how much total energy would be needed, which is 3.2 Tkwhrs. Since presently we get about 1.5 Tkwhr by renewables, that means that only 1.7 more Tkwhrs would be needed to get all energy by solar. That, again, would still be a heavy lift for solar alone at present, but it would be quite easy with solar, biomass, wind, possibly hydrogen, and the other future sources together.

Even in the near future, due to recent advances in making panels lighter, more efficient, cheaper, and more ecologically friendly, solar will provide considerable help and solace. For example, several new manufacturing plants in Europe and the U.S. are already constructed and will soon offer new solar panels made of abundant, cheap minerals like perovskite and silicon. These environmentally safer, more efficient, cheaper panels should soon become available and make a major impact on quickly increasing solar capacity.

D. Biomass

So now we ask, "How much energy could we get from biomass, and is this a feasible route to sustainably become the main future source of U.S. energy?"

Energy from biomass has actually long been forwarded by many scientists as probably the most sound and feasible way of 'mining' our energy (see, for example, John & Watson, 2007). Growing algae for fuel was suggested some years ago, but from an ecological standpoint (as well as economic and logistic standpoints), this option (i.e., using common combustion technology) does not seem very desirable, at least at large scale. Biowaste (sewage, wood, crop residue, and the like) is also often suggested. All these are under current interest, especially for small-scale operations. In fact, currently, there are already thousands of such biomass and/or waste electrical generating plants in this country and all around the world. Sweden, for example, gets 52% of its total energy from wood.

Most of the energy from wood and waste, however, is currently being produced by combustion, similar to most coal-fired plants. Burning wood, especially, and wastes, has been widely seen as being primae facie evidence for carbon neutrality: i.e., there would be no net increase in CO_2 in the atmosphere. Congress, in fact, in 2018, simply declared it so. A great many ecologists and others, however, disagree. In the short and intermediate time frame, calculations indicate that these processes are definitely not net zero. They are, of course, much closer than any fossil fuel but are still worryingly "carbon positive."

Unfortunately, by far the largest category of current "manufactured" biomass fuels are liquid biofuels in the form of bioethanol or biodiesel. Over 17 billion gallons of bioethanol are produced (by old, inefficient methods), mostly from corn on 36 million acres in the U.S. It, along with its' cousin, biodiesel, replaces over 10 percent of all gasoline for transportation. Biodiesel (70 million gallons annually) is produced mostly from soybeans, sunflowers, and canola, all raised on 17 million acres of cropland.

While the process of bioethanol and biodiesel production from plants is a relatively straightforward mechanical, chemical and/or fermentation-type extraction, it is also excessively energy-intensive to produce. Earlier, we had briefly noted that most ecologists and environmentalists see no overall ecological advantage to these fuels as derived by present agricultural means. Nevertheless, they are often, and loudly (and wrongly) touted as being, "of course, net carbon neutral".

Nevertheless, fuels derived by a new method from biomass are actually the most efficient and promising new energy source. 'Synfuels' are the newest form of biomass-derived fuels that are being increasingly used and considered for the eventual main sources of energy, even by the 'biggies' in the industry. These are the "real" cellulosic biofuels, as they are mostly made from whole plants. They can be readily derived from a great many types of biomass. They were first derived from coal and evaluated there as a more ecologically friendly way to use coal. However, it seems to most experts that biomass from living plants should be a better all-around source of energy than coal. Many studies (Lu et al. 2019) have shown that the atmospheric emissions, e.g., gases, particulates, tar, and other pollutants, are lower by the synfuel method than any other current or clearly foreseeable processes.

Yergin, too, believes that synfuels obtained through the gasification of biomass (e.g., synfuels or 'biogas') may have the best potential as the major fuel source of the future. Many other experts have also noted that energy sourced from agricultural biomass might be the best option going forward (Brandin et al. 2011). Much R&D and progress is being made in manufacturing the machinery, facilities, and equipment to more efficiently, economically, and ecologically-friendly produce quite large amounts of synfuel electricity directly. The technology is new and advancing rapidly. It is, though, already the most efficient production process in terms of energy conversion. Pahl (2007) notes that small-scale biogas generation has been used for many years, especially in converting landfills and other wastes to non-polluting use.

There are a lot of different types and methods and machinery for making synfuels. Basically, though, these are rather straightforward in their engineering. It consists mostly of a special furnace and associated gas-handling devices to handle the gaseous combustion products. Depending on the exact temperatures, pressures, and gases used (i.e., air, oxygen, or other), different products of the biomass are produced (and in different parts of the "furnace"). Most of the syngases produced from the plant molecules are combinations of CO, CO_2, methane, hydrogen, and water vapor, and the evolved mix is referred to as "producer gas". For our purposes, these biogases or synfuels are important components, as these can quite readily be collected and fed immediately and directly into a regular gas or diesel engine. In turn, these can run a traditional electrical generator. This process is currently used in a few of the most advanced coal-fired generators, too and seems to offer the best hope as the major source of our total energy needs in the future.

Noteworthy in this type of processing is the fact that unwanted emissions of all types, e.g., tar, gases, particulates, etc., are quite low. Gu and Bergman (2017) found that the net CO_2 emissions, too, were quite low. They reported that the CO_2 emissions per Kwhr equivalent was .33 from woody biomass, compared to 1.1 for coal and 0.72 for natural gas. This comes pretty close to carbon-neutral production and is actually quite efficient.

Another of the unique benefits of this kind of system is that it is a one-stop, compact, self-contained system for direct biomass-electricity conversion. Biomass from virtually any desired kind is fed into the furnace, and the final product comes out quite immediately, ready to convert to electricity. At the moment, these kinds of systems are still quite small and not yet ready for large-scale, grid-size introduction. Some, however, are ready for immediate small-scale use.

Perhaps the worst drawback of these generators, even the small ones, is that they currently lack intermittency. That is, they don't have the facile ability to have automatic, irregular OFF-ON cycles to 'even out' the energy usage. On this front, though, there is significant hope for great scientific advances in various arenas, as in battery technology, for example, to help solve the storage problem. There are already many new types of batteries available, including flow batteries and various kinds of all-carbon batteries, such as from graphene or possibly lignin, which should help.

So the question really becomes, "How could agriculture, using this new type of technology, be used to supply the bulk of our energy in the fairly near term?" The most logical plan would be for farmers to raise a special energy crop(s). Which crops? A goal would be to produce a crop(s) that is efficient, economical, and environmentally friendly, which can be harvested in a more or less traditional manner. These crops could be harvested directly and delivered for the production of cellulosic biofuel in the form of syngas. The most logical type of plant to use would be perennials, which have a low impact on soil, water, fertilizer, chemicals, etc., and high productivity. There aren't many plants currently under consideration that fit that bill. Switchgrass has often been suggested, as has been willows or other trees, and a little research is going on exploring some of these possibilities.

We will use, as an example of the whole scheme, a recent addition to the suggested biomass energy crop list, the tall perennial 'giant' grass, Miscanthus spp. This group of plants is widespread in Asia but has been shown at several Agricultural Research Stations in the U.S. to grow very well on almost all types of ground. Once established, it produces well for about 20 years. Miscanthus grows in marginal soil, requires little tilling, water, or fertilizer, and produces an average of about 7 tons per acre. Prof. Jeff Skousen of West Virginia University has shown that about 7-8 tons per acre can be grown without fertilizer, pesticides, or other chemicals on old reclaimed coal mines. It is the most efficient plant known in terms of its photosynthetic efficiency and rate of growth (it is a C4).

Both Prof. Skousen and a group at Iowa State University have been successful in growing and using Miscanthus for fueling renewable energy plants for the past several years. They find that because of its efficient burning characteristics, it has a low mineral and ash content of only 2-3%. Miscanthus has been used in the Iowa State University power plant for 5 years, and they plan to completely eliminate all coal by 2025.

Miscanthus, while its agronomy is not yet very well advanced (Christian, 2012), is already under commercial production for energy by a company in England and at a smaller scale in a couple of Scandinavian countries. Its burning efficiency for electricity is about the same as coal (about 33%), while the energy content of coal is one-third higher (6400 kwhr/ton) than Miscanthus (4720 kwhr/ton). This, however, is better than most other biomass materials, including wood, which is about the only major biomass material now used for small commercial heating or electrical production. It is also low in minerals, so it is less potentially harmful to the soil than most others that are commercially used. The ash can be used directly as a soil amendment. It is not only low in ash, about half that of most grasses or wood but also low in tar residue.

So, how could we use this crop to first test whether such crop systems might allow agriculture to totally replace coal? As mentioned, currently, the U.S. uses a little less than 700 million tons of coal. This produces about 25% of America's electricity (1.1 trillion KWh/year). So, to replace that 1.1 trillion Kilowatt-hrs with Miscanthus would comprise a sizable agricultural initiative. Since Miscanthus grass actually yields about 1600 kwhrs per ton, to replace the 1.1 trillion Kwhrs from coal would thus require 30% more tonnage of grass compared to coal. To replace that 700 million tons of coal equivalent would, therefore, require 910 million tons of Miscanthus. At 6 tons per acre, then, getting 930 million tons would require planting about 158 million acres. This is about a third of the total number of acres currently designated cropland in the U.S.

Obviously, it would not be feasible to just immediately transfer 150-plus million acres of land to produce the grass Miscanthus (or any other similar energy crop). Ideally, this transition would be slow and gradual and become a part of a (hopefully) more conservationist kind of (smaller) farming enterprise as a whole. A gradual switch to smaller-scale, more ecological agricultural practices would be highly desirable in any case. Hopefully, this would occur along with all of the other programs to gradually allow "re-naturalization" of the world, i.e., living sustainably and well with all the ecosystems. Aldo Leopold, almost a hundred years ago, may have said it best: "We abuse land because we regard it as a commodity belonging to us. When we see land as a community to which we belong, we may begin to use it with love and respect."

We will look, then, at how we might try a small test case to start with. If, say, over a decade or so, a few farmers in virtually every county in the U.S. were to assign maybe 25 acres to try to grow Miscanthus, that would make a worthwhile start. Just 25 acres would make a total of about 1.5 million acres nationwide. After an initial year of establishment, that would allow a harvest of nearly 10 million tons of energy fuel. That could be burned to obtain something close to about 1600 gigawatt-hours (GWh) per year of electricity. That, of course, would be only a small fraction of current coal electrical production but would represent a truly radical start.

One advantageous feature of using biomass for fuel is that small, local generating devices, like the biomass generators currently available, and, as envisioned by J. Rifkin and others, could become incorporated into more local use where operators would have a reliable, local source of raw materials. An individual farmer with 25 acres, with a harvest of 150 tons, would have a stable, worthwhile commercial crop with a reliable market.

How might this scenario work in a single County, say, for a group wanting to have a 30 Kw biomass synfuel generator for local use? The average U.S. household uses about 11 thousand Kwhr per year. Thus, a 30 Kw plant could supply at least 20-25 homes. Since Miscanthus produces about 1600 Kwhr per ton, such a generator would require about 150 tons per year. This could be produced from just 25-35 acres, from one farm essentially. Such a small local power plant could also be tied into an 'internet' grid (ala Rifkin) and also be a part of a State or national smart grid. (Recently, Gretchen Bakke ["The Grid"] nicely summarized the nature and history and problems as well as the future of this many-headed unseen monster. She gives us a good picture of how this rapidly growing, imperfect, evolving hydra might eventually be tamed to serve the multiple-sided needs of a future highly functional, efficient grid.)

In addition, all this offers the prospect to small farmers of having a perennial, steady cash crop that could readily be handled, processed, and used in the local area. Basically, only quite standard harvesting equipment is required for cutting, drying, and shipping. The sun-dried grass could be loosely baled in 'small squares', which would facilitate feeding into the small synfuel generators. This would offer great advantages for small users, including the local energy-generating operators. The Rural Electrical Cooperatives of the past, for example, might offer a possible good model for these types of distributed industries. Hawken (2017) and others have also suggested several other types of approaches that future agriculture could take towards the goal of providing both the food and energy crops sustainably.

Right now, a major hurdle for biomass and similar, local, distributed- energy- producing individuals or companies is that there is no market infrastructure to provide a reliable, cost-effective market for raw materials. Presently, in fact, this is a fatal flaw in any scheme to make energy locally available. The whole idea outlined would seem to be a viable and desirable soft path for major advances toward true energy independence.

Of course, a number of other biomass options besides Miscanthus would be feasible, too, based upon local sources, situations, etc. Locally or regionally, various other fuels, such as wood, crop residues, locally desirable grass crops, and the like, might be very advantageous as substitutes. As noted earlier, many workers have used and advocated for switchgrass. For example, Sokhansanj et al. (2009) have recommended the use of switchgrass for the same kind of potential uses. Switchgrass (also a C4 plant) is about as productive as Miscanthus and, while it burns at 5-10 percent less efficiency, is otherwise quite comparable to Miscanthus and could provide an excellent choice. Haiming et al. (2009) described excellent results with their synfuel plant with a 442 MW potential output using switchgrass.

In summary, this hypothetical system could be a new way to get a sustainable source of fuel for a goodly portion of our total future energy from agricultural biomass. Actually, we are proposing a revolution in energy. In a century or two, farmers could be the primary energy producers——and, of course, on a much smaller scale.

Chapter 8
Populations & Family Planning

A. Short History of Populations & Family Planning

In the history of civilizations, there have been few deliberate, widespread attempts to reduce population. China stands out as a major exception, although several other countries have recently done the same on a smaller scale. In the mid to late nineteen hundred, the Chinese government determined that the national welfare would be greatly enhanced over the long run if their population would be reduced, or at least stabilized. A strict one-child limit on parents was placed, which lasted many years.

No other particular methods of pregnancy prevention were promulgated, nor were any widespread planning or educational groups put in place. The parents, mostly, were responsible for figuring out how to achieve it under the pain of punishment. Chemical contraceptives and methodology were advertised to an extent, and abortions were not proscribed, but otherwise, not a lot of direct state effort was made under the broad heading of family planning assistance. It was, however, definitely taken to have a coercive intent.

Nevertheless, the initiative was quite successful. In 1980, when the plan was put out, China's population was (only!) about 922 million, but the growth rate was +1.35, and the population was growing fast. The number of births then exceeded deaths by 22 million per year. By 2016, when the law was changed to a two-child policy, the population growth rate had fallen to about + .3%, but the population was now 1.32 billion and, of course, still rising.

Then, perversely, the Chinese deemed the initial project too successful, as their slowing and seemingly stagnating, aging population began to worry them about future economic growth. Hence the two-child policy. The growth rate was still slowing by that time, to just above zero. The projections, though, were to go negative, to -

.5 percent by 2050. This would have led to a population at the 1990 level (1.117 billion) by 2050. However, in January 2019, the Chinese Academy of Social Sciences even issued a dire report on the situation, and the government cancelled the two-child policy. They warn of a looming population "shortage" and an economic crisis due to their current aging and falling population, with a predicted shortage of workers. What the next China reaction and policy will be is of great interest to all of us. (It is necessary, too, to emphatically state that the population of China is way above a sustainable level—along with most other countries).

It is noteworthy that India has surpassed China as the world's most populous country. India's growth rate is now about 1 percent. While the rate is decreasing slightly, India's current population of 1.4 billion will continue to increase for many years. Weisman (2017) noted that estimates for India are that their population will peak at 1.68 billion by 2055 and then (maybe?) start falling. In India's most populous State (Uttar Pradesh), he reported that girls who had started into secondary education had an average of 1.9 children, while those who graduated had 1.6. The (majority) of girls who had no secondary education at all had an average of 6.0. These statistics highlight a significant fact for family planning and governments all over the world. India has had no really discernable population policy, but general contraception materials and information are increasingly available to many women.

It is also well to recall that the world population is still growing at 1.1%; if this rate were to continue, it would mean a doubling of population in 63 years. (If that were to actually happen, yes, that would mean a world population of around 15 billion by 2085!) While it is true that the birth rate worldwide is falling, it is still a healthily growing 17.87 births per 1,000 people. The average TFR (total fertility rate) is still 2.4, albeit this is down considerably from 4.7 in 1950. Unfortunately, there are still 60 countries that currently have a TFR over 3.0 and 95 total countries with a TFR over the replacement rate (2.1).]

South Korea is another country that has deliberately pursued a population stabilization plan. In fact, they now have the lowest birth rate in the world. Their population today is 51, 610,000 and their growth rate is essentially zero. Right after it became South Korea in 1948, its population of 40,500,000 began to grow tremendously. For several years, the birth rate was 6.1 per woman, and growth rate was

2.8 percent. By 1963, this had begun to put a great social and economic strain on the country, and the government started a vigorous national family planning program to ease growth. They distributed free birth control devices and information, encouraged parents to have only one child, and had classes for women on family planning methods.

These programs worked. By 1983, the birth rates were falling (to about 2.4 births/woman), but the population was still expanding too fast. The government then stepped up its population control measures to include granting subsidies if a person agreed to be sterilized. Also, they began suspending educational insurance benefits for maternal care if the mother already had 3 or more children. Later, they denied tax deductions for education if the pregnant woman already had 2 or more children. The program worked very well. Too well, it seems to many of them, as they are now worrying again about severe economic strains, this time because of a lack of workers in the shrinking young population. Sound familiar?

Nevertheless, the birth and growth rates in South Korea have continued to fall, and today the births per woman is 1.21, among the lowest in the world. Today, their population is 51,611,000, but due to the so-called 'population momentum effect' is expected to peak at 53 million by 2031. Then, the population will begin to fall, and by 2050, they will wind up with a population of about 50,595,000. Not a very remarkable endpoint, considering the urgency of the problem. But still, it shows the power and efficacy of standard family planning and population-slowing methods.

However, even though South Korea's population is almost static at the moment, because of the current relatively large older population cohort, they are already worrying about their present economic, especially worker, problem. It feels to them like their population stasis presents an immediate threat and a potentially permanent worker problem. This is the familiar refrain (like in China's demographic crisis), and some way must be invented soon to make it possible for countries to get over these population reduction blues.

Notwithstanding their current official blues, it is an interesting fact that today, and for many years more, there will be 530 people per square kilometer in South Korea. That is crowded by any standard. The population density of Great Britain, the most crowded European country, for comparison, is 345, and Germany 235 per square kilometer. For further comparison, Canada has 3 persons per square kilometer while Italy has 197. Worldwide, the average is 14 people per square kilometer of land (which is really a lot); the world is a crowded and frantic place.

The population trajectory of Japan has been quite similar to South Korea. For example, there were 1.9 million live births in Japan in 1970 but only 977,00 in 2016. Prior to 1970, Japan's population grew very fast and in 1974 numbered 111,900,000. That was up from 90 million just 20 years earlier. But since then, growth has slowed, and by 2018, the population had actually declined from 128,240,000 to 126.8 million. Their birth rate dipped below replacement level over 25 years ago. In 2019, there were 400,000 more deaths than births (the world's fastest population decrease). The population is expected to continue to fall quite fast, to 'only' 100 million by 2050 and 85 million by 2100.

The exact reason for Japan's impending loss of population is not entirely clear but most probably due in large part to the general lowering of fertility by "choiceful" means, a 'by-product' of the rising economic status—— the so-called industrialization effect. But there was no overall, designed, definite government plan or campaign in place. Early on, there was an effective effort to increase the availability of information on standard family planning, contraception, etc. This was helped along by a general sensed desire to slow the population growth. In any case, the choices were made family by family, sometimes by design, often subconscious. But whatever was the cause, it worked. This is expected to be the result in every situation where there is a general will, and women are empowered, educated and allowed a choice. This is the 'secret' to the Velvet Touch. (This is also why almost half of the countries of the world have a reproductive rate that will bring population growth down—— and below zero if they keep it up.)

Are the Japanese happy with the result? Quite the contrary; just like in South Korea, the governmental and popular response to the situation was (is) to decry the development. They worry now about the economic problems already being experienced, and they fear massive problems further down the road. Japan's population is aging very rapidly, too, and they fear running short of workers and pension funds. (They have the highest average age of any country in the world at 46.3 years.) Significantly, however, the relatively fewer young people do not naturally desire as many children as their predecessors.

The only official program response in Japan recently has been to increase the birth rate! And apparently, no policy initiatives will be forthcoming in the near decade regarding further population reduction either, except possibly, and unfortunately, to augment their already pro-birth suggestions. Fortunately for us all (and eventually for them, too), there is no indication that Japanese women will change their fertility rate much, so the population will probably decrease even faster than the projections for a time.

Wiesman (2017), however, also noted that many people in Japan are actually starting to think about how to more effectively deal with the realities of a non-growing population in regard to economics and social and governmental policies. He quotes Professor Akihiko Matsutani of the National Graduate Institute for Policy Studies: "Paradoxically, our shrinking situation could end up being beneficial. We have to change our business model. That usually takes a long time, but we can't wait. This is the moment we have to change."

A similar situation was, and is, developing in many other countries too. For example, Italy now has a static population of about 60 million, but its growth rate is slightly negative. Thus, their population is dropping a bit. There are no active policies or intentions of the government to reduce the population; it is just happening, as is usually the case where women have real choices.

Leaders in Italy, too, have started decrying a "reverse population time bomb" to describe the problem of population stagnation that arose. In this case, they lament problems, especially with worker shortage. Equally importantly, they fear being unable to pay for the aging population's medical, educational and retirement expenses. They worry that the economic problems encountered are worse than the original crowding problem. The government is responding by (mildly) encouraging immigrant workers, but otherwise, there are no real or effective policies being prioritized either way. Sadly, the government is now trying to stimulate the birth rate, offering monthly cash bonuses for pregnancies ("baby bonus"). In the meantime, of course, the air pollution situation, especially in the Po River valley (which is rapidly drying up these days), plus several other metropolitan areas, is increasingly worrisome.

The same kind of situation is occurring in several other countries, too, of which Spain is a good example. Spain has a TFR of only 1.26 and has already lost 400,000 people since 2012, so they are now "officially" already in the population reduction phase. (As noted earlier, this was accomplished in a great many other places, too, without any official governmental policy or even conscious awareness, just the result of millions of individual choices and, importantly, an available contraceptive smorgasbord.) However, the current Spanish government is now apparently livid over this situation and is actively seeking programs to reverse this unwelcome trend. They have recently appointed a "sex tsar" to develop programs to increase the population.

Here we go again. Let's see: during steadily growing populations, an economic crisis, like in China or South. Korea was feared. Then, when population growth slows or stagnates, that, too is an economic crisis. It seems like there may be something inside the socio-economic system that needs fixing in both cases rather than trying to make the population numbers create and/or fix problems.

It is discouraging to see this knee-jerk, common response to what could maybe best be seen as a much better solution, i.e., stabilizing or reducing the population and revising economic strategies. It is hard to see how adding more babies with the aim of fixing and maintaining an economy in a country that is already overcrowded and environmentally overtaxed can be a useful answer for very long, or at all. That simply adds to the problem, not part of the solution. And to rapid, further degradation of land, air, and water, i.e., Nature.

Besides China, which used both persuasive and coercive means, many other countries have also recently drastically slowed population growth by various means. Some of them are already at zero replacement rate, i.e., 2.1 children per woman. Among these are Brazil, Iran, Mexico, Sri Lanka and Thailand. They have gotten at or near replacement level in just a decade or two of government-backed policies and readily accessible, affordable family planning services. That's all that's needed.

On cue, though, in many of these countries, there is growing concerned among some economists and politicians about the coming 'problem' of an aging population, lack of workers, and potentially slowing economic growth! Buyer's remorse and cold feet are common maladies when it comes to population. Even Brazil, Iran and Mexico's population reduction initiatives have been slowed by these economic maladies recently. (We maybe should call the "population reduction blues" the "dollar bill blues" since it removes the main driving force of all economic systems——growth.) This is, in fact, our fatal disease: growth addiction. This is the ultimate disease on which therapy must begin.

Iran has had a rather roller-coaster dance with their population, too, over the last half-century. From 1960 to 1979, under the Shah regime, some attention was paid to slowing their population growth. Their population rose from 22 million to 37 million during this period. After the Revolution in 1979, however, Ayatollah Khomeini wanted Iranian women to birth a 20-million-man Army. Naturally, the population rose to almost 55 million by 1989, when he died.

The next Iranian government, though, instituted a large and effective family planning program using the standard methods that could stand as a shining example for all countries. It was successful, and the fertility rate fell to 2.0 in about 20 years. Then, the current regressive policies were re-instituted by first Ahmedinejad, then Ayatollah Khameini. Free contraceptives, i.e. OC's, condoms, IUDs, etc., were cut off, so the population again is rising, although not quite so fast. Their current population is 83 million and rising, but not so rapidly anymore. This 'accidental' brake on population growth seems to be a clear 'holdover' of fertility changes in women from their previous experience with having family planning and smaller families.

In so many countries, though, we have the complete opposite trend from South Korea, Japan, Italy and similar. Nations where the population is still totally out of control, and the people (some of them) cry out for rescue. Uganda is one such country. It has the highest fertility rate in the world. The population is about 40 million now but projected to be 90 million by 2060. It has welcomed some family planning, although there has been little governmental policy or programs. Several International agencies have been working in the country to set up many local (effective) programs, although these efforts have had little country-wide effect. Recently, funding from the international groups to provide free IUDs, condoms, OCs and the like has actually declined dramatically. The population is now essentially exploding. Wiesman (2017) said that at least 46 % of Ugandan women lack access to any form of contraception. Like in many countries, social norms and traditions (unfortunately mostly set by men) have traditionally demanded many children. Along with the population explosion and squeeze on the land, the country is now down to about 100 live chimpanzees.

Neighboring Tanzania is also in the middle of a disaster-bound tailspin. In 1900, Tanzania, only about twice the size of California, had a population of 250,000. By 1950, there were over 7 million, with a Total Fertility Rate of 6.9. Not surprisingly, by 2012, there were 47 million people. Today, there are 59.7 million, and their growth rate is still 3.0%. They estimate their population by 2050 to be 100 million and 250 million by 2100. What will be left by then of their iconic Gombe district (and everything else) is not hard to predict. Their neighbor Kenya, with their iconic African vistas and wildlife, is faring about the same with its impending population and environmental disasters.

In view of all these kinds of data, we might all, like the journalist Declan Walsh in his new book "The Nine Lives of Pakistan" say, "The most pertinent question might be not whether Pakistan will fail, but how it has survived this long." This could equally well be said about the dozens of other overcrowded places around the world, like Bangladesh, Rwanda, or Guatemala, for example.

It seems almost like a sick joke that just having the population leveling off but still keeping a high population density would produce such angst. One might think, rather, that it might offer some hint of future benefits and continued prosperity. Again, let's see if we got this straight: when their population is too high, the country scrambles desperately to create more jobs and also worries about unemployment and other economic disruptions. Then, when population growth is decreasing, the cry goes out even more vehemently that the lack of workers and non-growth is causing massive economic problems. Low growth or declining populations seem to them as presenting existential economic problems. In this merry-go-round, you can't win.

However, it should become obvious that a declining population over a longer term will surely help solve, at least to a considerable degree, the humongous problems with the "life-o-sphere" of any given overcrowded nation. We must quickly find ways to defeat this 'no-growth blues' situation. It seems like an idiosyncratic situation where the cure is considered worse than the disease. The socio-economic system would seem to be the problem, not the other way around. The tail should not wag the dog.

B. What Should We Do About Growth?

This whole thing is all backwards! If you can't stand for three seconds a non-growing population with which to maintain a totally growth-driven economy, then there must be something wrong with the system to start with. The past three century's magic, seemingly hard-wired-in-mode-of-thought, is now not only old-fashioned but clearly dangerous. It is past time to start seriously making a new "Human Game" and a theory and practice guidebook for States and economics for the next centuries. That shouldn't be all that impossibly hard. Unfortunately, it would seem that nobody (certainly no Nation or State) has really seriously tried to work at inventing the steady-state economy. The will to start seriously working on it is the hard part.

What we expect, of course, is some of the loudest, hardest wailing known in political history over the outrageousness of the necessity to get off the ever-upward arc of the historical limitless growth mania. Most people just unconsciously thought that people who were optimistic about the future and acted accordingly were 'making progress' and 'doing good'— "naturally". The past few centuries of manic growth have actually served humanity well in many ways, but it is time now to think our way out of this ethic, which is no longer working for the natural world—and soon enough for us, either. We need to provide a sounder basis to keep the human enterprise going on a healthier path. The disease of Growth Mania must be stayed. It seems, however, that in just about every country, the stance is to stick with the 'Goodie train' that has always seemed to work for them; that is, their growing economy must remain the engine of a perpetual growth and progress machine.

Bill Gross, a financial analyst (2019), is one of many others who let the cat out of the bag: the dirty little secret that any of our economics do not work unless we get either a growing population or plenty of immigrants. Obviously, depending on immigrants to supply the workforce is an outlandish idea, as these come as supernumeraries from somewhere else with overgrown populations. Bricker and Ibbitson (2019) said it out loud: the feeling among economists, politicians, businesses and others that "accepting immigrants isn't a question of compassion or tolerance. It's good for business. It grows the economy."

"In other words, swell all the ills of overshooting carrying capacity, to ensure more human capital, more cogs in our country's economic machine."

–Dave Foreman

Wealth (and power and associated personal 'bennies'), of course, has become virtually synonymous with growth. Wealth, as Pinker (2019) and many others have shown, has been increasing rapidly all over the now thoroughly globalized world. It is being created not just by digging it arduously out of the ground but by human capital and innovation. It is increasingly being made 'immaterially' to a considerable extent, that is, by digital or information mining. But still, all based on the growth of people, markets and resources. Hopefully, with a somewhat different perspective on Nature and non-material things, there will be a lessening of the need for lots and lots of ancillary materials and resources (e.g., Palmer, 2020). This would help alleviate some of the calls for hard-core, full-time, hard-labor pursuits of the traditional types. Maybe, as Hazel Henderson (2007) said, "there could be alternate ways of making money work for people, rather than the other way around."

Lots of suggestions and ideas have been floated around in the past couple of decades, both by a few economists, political scientists and others, to at least give some positive grounds to believe that we, the people, can invent new ways and solve the transition problems without catastrophes. Blewett & Cunningham, 2014; Reich, 2018; Daly, 1977, 2007; Jackson, 2017; Hedges, 2018; Alpervitz, 2011; Rich, 2019; Heinberg, 2021 and many others have made some progress in developing new approaches to government and economics, especially aiming at the main disease of the ultimate, primal need for fast and continuous economic growth. It seems, too (Bregman, 2017; Banerjee & Duflo, 2020, Lowery, 2018), and others) that some sort of guaranteed basic income system will have to be a part of the new economics since hardcore jobs will likely continue to diminish for a while.

As seen, a shortage of labor is one of the earliest howls of anguish coming from the corporate and economic world, even as populations just stabilize, let alone decline. Dr. John McKee, Dave Foreman and others have suggested that raising the retirement age would seem to be one non-onerous thing to do to help that to some degree. McKee says that a great many old folks could continue to contribute in many specific arenas well into their eighties if needed. Increasingly, in the coming new world, the jobs needed will be in information, education, health and medicine, leisure, travel and entertainment and non-manufacturing jobs.

However, a continuance of overpopulation just because of purely economic considerations, or how to get enough workers, or pay for retirement benefits and the like is the stupidest and most shortsighted proscription for the future centuries one might imagine. As tallied in previous chapters, how the enormous growth and size of the human population has driven the biosphere to its breaking point makes it imperative that even previously unthinkable options for remedies should be advanced.

Of course, reducing the population alone will not solve all the pressing economic and environmental problems we have. It should, however, make the tasks of ameliorating the present perceived problems more manageable and provide additional psychological stimulus for keeping up the good fight. (And it will, in the end, wind up saving the living planet!) Klein (2014) noted that "to arrive at dystopia, all we need to do is keep barreling along down the same road we are on using our reigning economic paradigm."

It should be obvious that at the present level of populations, trying to raise the standard of living for most of the earth's peoples to that of Norway, say, will be extraordinarily difficult (nay, impossible by the standard means) even though urgently desired. It is certain that continued population growth, or even stasis, will mean relegating the already poor countries to static or even lower standards of living. Equality is much-mouthed, but no serious governmental steps anywhere are contemplated to permanently solve this problem. It seems inhumane and unacceptable to knowingly and willingly consign a good part of the present and future human population to increasingly permanent misery levels, much of which was itself generated by overpopulation.

The old bromide: 'Of course, we have to grow the economy of our Town/ Country/State/County/Company,'— becomes, at last, simply a stupid lie.

Trying to get pure scientific or technological fixes to the increasingly serious environmental problems is also a fool's errand. It is indeed a good idea to try to raise standards of living and to develop more efficient ways of providing fully for needs in developed and under-developed countries. The problem is that there is no way to square the circle. That is, there is no way that the Earth, either by way of increasing food or energy production, could accommodate the great expansions and strains that would be necessary to accomplish this on our current trajectory. There can be no workaround by Big Water, Big Energy, or Big Science to reduce the need for population reduction.

Even if we tried, the strains and further catastrophic damage that would be done to the natural world would leave everybody in an increasingly dysphoric, ultimately fatal, situation. Mindless growth in service of money (= Power) or supposedly limitless resources has been the mantra of the past several centuries. Moynihan, 2020; Crary, 2020 and others have said it loud and clear: Mankind, for some time, has been mindlessly careening towards extinction with our grabbing of every inch of the planet. Franklin Roosevelt knew it already in 1932 when he said, "Prioritizing property rights, business interests, and the markets is detrimental to human freedom."

Putnam (2020) says that in the future we will need to create a civilization that has a better balance between the "I" (which is now predominant) and the "We". He says this balance has been going in the wrong direction in the U.S. since the 1960s. He also said that we had earlier (about 1905) been on an "Upswing," and we could do it again.

Chapter 9
Population Reduction Plan

A. Plea for a Cure

OK, here is the basic proposition:
(a) since all indications are that we are way over the carrying capacity of the earth, and thus, exceedingly bad disasters will befall the whole world if we don't make drastic changes; therefore,

(b) we need now to develop general plans worldwide for how to go about population reduction and also changing some of the rules of the 'Human Game.'

The United Nations, private, quasi-government and other volunteer groups have long been engaged in many initiatives to advance family planning in individual countries or areas. The Gates Foundation alone has spent probably nearly a billion dollars in their long efforts to supply information and materials to families, communities and areas. The impact of all these efforts, while sometimes locally powerful, is largely unknown and generally only locally successful. Never, though, has any government yet even advanced the notion that their country should make any attempt at seriously permanently <u>reducing</u> their population. The goal, when there has been any conscious effort, has been to maybe slow runaway population growth. Unfortunately, the idea of purposely aiming to reduce the population greatly is seen as a preposterous and mad, ridiculous idea by most. We need to start changing this mindset and perspective over the next few years. Cultural evolution is urgently needed by wise people now more than ever.

The Union of Concerned Scientists issued a pamphlet back in 2002, Warning to Humanity.
"If we don't halt population growth with justice and compassion,

it will be done for us by nature, brutally and without pity—and leave a ravaged world."

Many scientists and others have made a near air-tight case for how and why the environment and entire biosphere is getting into a dire condition as a result of the population explosion. And also how we are heading for ultimate disaster if solutions are not instituted quickly and thoroughly. One of these enormous projects needing to be dealt with, of course, is global warming. Many countries are already steering in that direction and adopting policies and practices to begin to achieve the goal of reducing the dose of carbon dioxide in the atmosphere. A gigantic data base of information, technologies, industries and groups ready, able and willing to accomplish that task are within sight. Wire (2009) and Murtaugh & Schlax (2009), however, perspicaciously showed that "family planning is far more effective for climate change intervention than green technologies."

The highly positive and quite fast success seen as a result of governmental and/or societal desire to reduce population growth in some of the various countries that we mentioned earlier reinforces the widely held, and actually true, notion that women really usually do not want large families. They are very ready to have small families, which is the sole requirement to stabilize or reduce the population. Several recent studies in various countries have confirmed this reassuring fact of life.

Of key importance under the new (hopeful) condition of falling populations, unprecedented in written history, a new economy and various other social structures will have to be developed. It must be one that is not based on the old verity of growth and its adherent mad financialization and market mania. The addiction to limitless growth must be healed. That would mean, for example, that every store or company will no longer have to have new customers, higher sales, and higher production every year to maintain jobs or their livelihood. Land speculators, too, and thousands of other groups would be hard hit. Farmers and landowners would no longer find it easy to, at any time, subdivide their land into lots and eventually make a million from eager new potential buyers who are being added to the roster. Of course, all kinds of people would need to be helped out during this long transition period.

GDP (the holiest of links to prosperity) "must grow by 3 percent per year" is an outmoded shibboleth, now almost a disease to be rooted out. The long-standing holy grail of most governments fueled by the natural growth of peoples and their demand for more and more goods, money, etc., has to be replaced with something more sane. Certainly, it must not be based on financial markets and the rising population, like it has largely been in the past. The prevailing assumption of continuing, unending material growth in a finite world has become untenable, impossible, idiotic.

The heart of economics today is about the ultimate efficiency of the Free Market. 'If governments don't meddle, it will always work out fine—as long as you can have growth. 'The Market knows best'. "Government is the problem," according to Ronald Reagan. As Heinberg paraphrases this philosophy, "Nature is merely a subset of the economy—an endless pile of resources to be transformed into wealth and keep growth going." He adds that growth needs to be redefined in terms of better, not more.

While conservatives assume that the market is both natural and effective and should have primacy over the way we should arrange our lives Konzcal (2021) says we should instead strive for more freedom from markets. Bregman (2017), McKibben, and others, e.g. Hawken (2017), Reich (2020), have also presented many new ways of thinking and doing business. McKibben's new book (2019) powerfully lays out how the 'Human Game' we've played over the past many centuries, where capitalism has captured America, and corporations have captured democracy, is now about played out. At the least, it has led us to the edge of catastrophe, where a life of meaning, purpose and dignity will become ever more difficult or impossible.

And no, the market does not always know best. Chris Hedges (2018), in fact, says, "Corporate capitalism cannot be reformed. It only knows one word——more. The Corporate state, however, is in trouble. All the promises of "free markets", globalization and trickle-down economics have been exposed as an empty ideology used to satiate greed".

But we should always remember Bartletts' Axiom:

"Population growth and/or growth of consumption of resources cannot be sustained."

Period!

As mentioned earlier, many countries, regardless of their population status, have started more-or-less continual worrying already. They worry about the wrong things, though. They pre-worry about slow or no growth, unemployment and under-employment, increasing use of robots (which replace workers) and the like. There would seem to be a common (Gordian Knotted) thread here. If some worry about not enough workers, others worry about too many workers for too few jobs due to robots, etc. There should be available some kind of happy medium—perhaps a permanent fix to this persistent merry-go-round. Basically, it seems that could be primarily done by a thorough rework of the prevailing mindless topsy economic system. 'It's the economy, stupid' is a splendid political ploy, but we say it's the stupid mindset that will cause us so much damage, and IT needs thorough reworking.

Ironically, with decreasing populations, it would seem necessary to increasingly call on more 'robotization' in industry and commerce to not only displace a lot of the drudgery work but also to maintain a desired level of stable production. Somewhere in there would seem to lie at least a partial fix. In any case, solving these kinds of gigantic and difficult problems has to involve a plethora of new ideas, policies and economic and governmental thinking. To avoid even bigger looming problems, attention to all aspects of government and society, as well as the best application of new or anticipated technologies or science, is called for. Prosperity through growth, as Jackson so elegantly lays out, is no longer conscionable or possible. As his book says, it needs to be replaced by prosperity without growth. This would at least give humanity a chance at avoiding premature extinction.

Reduction in population wouldn't be all bad, all pain and suffering. In fact, the feeling of letting some of the air out of the pressure cooker of constant growth, o'er-weening materialism, and incessant bombardment from all sides should be refreshing, even after a short while. If populations were to start reducing, the overall burdens on society and the environment should be decreasing, all the while making the other major problems progressively a little easier. Declining demand for food, water, energy and materials, for example, should be seen as an aid and balm to the other global, painful problems that we have.

Using our centuries of accumulated knowledge for continuing the pursuit of happiness, it should become easier to produce more satisfaction and well-being, one would think. The landscape will be more satisfying to look at and play in. Less hassle and traffic and crowds at the beach and Parks and streets and everywhere in between. Less stress and anxiety. Every decade won't bring new subdivisions or manufacturing parks or dizzying expansion of new developments to have to rapidly adjust to. Everyone in the U.S. is all too familiar with the difficulties in assimilating the dizzying changes in our towns and neighborhoods by the influx of new peoples and new streets and industrial parks where once their uncle's silo stood. This kind of thing is a significant factor in the rising tensions, havoc and anxieties generated in our teeming and increasingly creaky world.

And humans are clever monkeys. They could figure it out. It is in our own wonderfully adaptable human nature. (Fortunately, too, it is a fact that our species did not evolve as a conspicuously selfish one.) There will be a little time to give the governments and societies of the future enough opportunity to develop more humane and satisfying human ecosystems as well as social, economic and governmental arrangements. An ecosystem does not depend on any 'economy' for its survival, except if that economy also has a death wish. Economics is purely a mental construct of man, not a universal fact. Life really isn't all about the economy. Gifford Pinchot said in 1912: "Life is something more than a matter of business."

Some countries with zero or below population growth at present are already dealing with the social and economic repercussions associated with their new no-growth conditions. Hopefully, some of them will be among the leaders in developing some effective new solutions. The process of solving the technical, social and biospheric problems and simultaneously carrying out the proposed low-key, painless, unseen, below-the-radar prescription for slow population reduction should really be something that gives hope and solace to peoples.

There has never been a greater crisis for humanity and all the world, nor a greater need for heroic efforts, so the response needs to be commensurate with the need. The payoff will ultimately be an increase in human well-being and a more pleasing and continuing natural world. A world that works not just for people's (especially the well-off) benefit but for animals and plants and all of nature. Singlemindedly striving toward, as so many are doing now, even if semi-consciously, a world of Man, by Man and for Man only, should now be seen as a myopic, disgusting worldview. Paul Collins (2020) puts it this way: "In light of the assaults on biodiversity, it is hard to maintain the notion that so much of human activity is rational, when in fact, it is utterly irrational, with our economics posited on endless, infinite growth in a world of finite resources."

The journey we are advocating here seems to us to be the only acceptable approach; that is, to tackle the multitude of problems at its source (at least start on them). It may last a thousand years. Never in the history of the world have governments or societies agreed upon and actually broadly implemented overarching plans, programs or policies to guide a country along any singular, one-track path.

Actually, Wait! What? That last part doesn't sound quite right! In fact, just the opposite has been true. Since the seventeenth or eighteenth century at least, it has been the modern, unspoken, singular, semiconscious, universal policy of just about every country, every State, that every man expand their land, their industry, their wealth, their community, everything. Growth in population, industry and material progress was the norm and the expectation. The world was Man's property, and they turned it all to their profit. Altogether, it was assumed that all should be engaged in becoming more wealthy and powerful, if not healthier, wealthier and wiser. All driven lately by unbridled free economic capitalism, fueled, at the core, in great measure by growth—and aided by selfishness, short-sightedness and greed.

And that program has worked—— almost to perfection! We won! But that will last for just a little while longer. We have grown ourselves almost out of a decent house and home. It is time now to get wiser and build on what we have, to begin to devise and invent and grow a smaller (actually, larger-horizon), happier and better world. Our own generations and preceding ones got us (mostly accidentally) in a one-of-a-kind Goldilocks era, into our current high status and also into our present serious conundrum. Now it is time for current and coming generations to help us dig out (or at least not dig us further in). It is meet and right to admit here and now that much of what we're doing now is a mistake, and now it's time to correct ourselves and start along a previously unthinkable, sustainable path. It is getting painfully clear that our survival is at stake.

B. How Much Should the Population Be Reduced?

How many people could the earth safely support? A group at the University of Texas (Pianki, 2006) stated that Americans use about 23 acres each. If everybody on earth did the same, then we would need 9-10 earths to support everybody. That means that the earth could support forever only around 800 million people. On the face of it, then, that number would seem to be about the maximum safe, human- and nature-conserving carrying capacity. Even at that, they suggest, we might be living on the edge. At least then, though, with some human engineering, some things could be improved upon, and it might last.

Joel Cohen, among a few others (e.g., Crist, 2012), over 20 years ago called for population reduction down to about 1- 2 billion. He, among others, says that number would be about the carrying capacity of the Earth on a continuing, safe and satisfying sustainable basis. He also called attention to the fact that the carrying capacity of the Earth depends on what level of production and what lifestyle people will accept. His own suggestion was for, eventually, one billion. That would at least theoretically allow a decent type of living condition for all people and (possibly) also maintain the natural world at a reasonably pleasant and sustainable level. A place where all people (and plants and animals, too) have "enough". At these levels, people could theoretically still have some of the good 'stuff' they have now and still live a good life. There is no exact number of people to fit the earth and Nature, just not too many.

So, again, how much would we suggest population, within the next century or two, be reduced? Most estimates of safe and sane human carrying capacity, assuming a quality of life about the same as an average European, hover around this one billion mark. Since there is no fixed number, we, too, are suggesting, like Joel Cohen, to at least temporarily aim for around one billion. This is partly based on a statement by Alexander et al. (2016); they say that if all of the current world's population had the American diet, it would require ALL of earth's habitable land, plus 38 percent more to feed them all. That, too, calculates out to having a world population of around 800 million people. Paul Ehrlich and colleagues (1994) also ciphered that 1-2 billion might be near the safe maximum. This is about the same number of people that existed in the nineteenth century. Paul Collins (2021) has also recently powerfully argued for a similar goal, one that is below Earth's minimum safe carrying capacity. Roderick Nash

(2012) also has suggested that 1.5 billion might be about right. He says, "Only limited numbers of humans can enjoy unlimited opportunities. So we will need fewer people and more advanced technologies and ideas."

As Crist, Foreman, and so many others so often say vis a vis what is the human carrying capacity,

"it is not the core issue to finely test the real-world limits for food and space and fiber, but what kind of real-world we want to live in."

If, then, the population would be restored to someplace close to one billion and to equilibrium with the rest of nature, then drastically reducing the individual ecological footprint would no longer be such an imperative. Possibly, but hopefully, a more mindful appreciation of our place in nature would drive some more environmentally (and socially) sound practices all round (like more recycling etc.). The first big hurdle, though, is to determine that the population ought to be reduced to allow a better life for all inhabitants, plant, animal and human, in the coming centuries.

Dave Foreman (2014) gives some guidelines on population reduction and also powerful reasons for doing so, saying: "We have a couple centuries, maybe. This is all predicated on someone, somewhere, able to make the world aware of the problem .... Population reduction is the kind of term that puts economists and politicians in fits of denial, for their system of creating wealth and building a power base depends on sustained growth."

The 90+ countries with TFR below 2.1, of course, have already essentially done much of the initial work by the standard means. These need only to be continued for a long time and also become widely adopted around the world. This will require a conscious decision made by societies and governments to do this. This also means, most critically, an overhaul of ways of thinking about economies, governance, etc. and even how society and civilizations should work, etc.

Robert Engelman (2012), one of the country's leading experts on population and family planning, points out many salient facts that will need to be incorporated into the kind of population reduction project outlined here. They report that nearly half of all pregnancies are unplanned and unwanted. Also, only 55% of the world's women have access to or knowledge of modern contraceptives. Presently, 44%, or about 2.3 billion women of reproductive age, use no contraception at all (Reyerson, 2012). Those are the reasons we have had the population explosion over the past many decades. Nothing illustrates the nature of the core problem better than that datum.

C. How to Do It

So the plan is by now plain. The world must start on a conscious, centuries-long journey aimed at seriously reducing the human population to a sane and sustainable level. The exact number of the population target level is not critical at this point. But the goal is unshakable—— the carrying capacity for humans must be generously sustainable, with ample room for natural ecosystems and for fuller and deeper expression of human well-being (beyond just "making a living"). In other words, Man's ecological footprint is too big. It urgently needs to be reduced to result in fewer people to make footprints. The basic idea of this book is to use, when and as needed, the standard birth control (family planning) methodology essentially as currently practiced. The overarching impetus and major emphasis for all this is simply to conserve ecosystems, men, beasts and everything in between, now and for the future. (And not just for the future of the next 50 years; no, we mean hundreds of thousands of years.)

Doing the fixing of our developing crises of Nature will be difficult, ineffably difficult, perhaps impossible. But it is absolutely necessary to try to save the Earth from losing much of its ecosystem support system and a big part of plants and animals, not necessarily excluding Man. Lots of technology, old and new, will undoubtedly be needed at every point, but technology alone will neither cure the problem nor assure future survival, tranquility or happiness.

Doing the technical part of reducing the population should, in theory, really be quite easy. As noted earlier, many countries have done it already (albeit half-heartedly, sporadically and sometimes almost accidentally). The hardest part of all is getting the will to even begin. The plan is simply to continue family planning in each nation, all around the world, for generations until a more reasonable and agreeable number of humans reside sustainably and well on the earth. (By the way, no, we are not now, or ever, in any danger of 'running out of people' or getting to an empty planet as a result of our present or proposed reduction.)

One handy characteristic of the solution is that it is not technically difficult, complicated, or even expensive. Essentially, all of the science needed has already been done. It doesn't require a large force of high-tech personnel or equipment. All the physical agents, equipment, procedures and all the rest are already in place in most locations around the world. Medical expertise, facilities, technical support systems, clinics, and all the materials, e.g., contraceptives, etc., are already available, or soon could be, around the globe. A slim majority of people are, in fact, already generally aware of and using the materials, techniques, or equipment that are in general use for family planning and contraception anywhere around the world. These are the tools that are causing the reduction in birth rates seen in nearly half of the countries already and are all the tools we will ever need. (In the U.S., as well as much of Europe and many other nations, no additional definite program really would be needed at all since their TFR is at or below 2.1.)

This effort should be seen as parallel to any other concerted action aimed at solving, ameliorating and grappling with all the overarching problems afflicting us all today. But still, the main urgent need is to apply the already existing population reduction techniques to the rapidly expanding countries immediately.

It sounds too easy, you may be thinking. But for many, the first critical factor, and the hardest, may be to remove the siege on those very people, organizations and bureaus that are already carrying out the good work. For decades, the clinics, research, organizational structure, funding and especially the people who man the fertility and other clinics have been under heavy siege from all sorts of enemies. Christian and other fundamentalists, pro-growth devotees (these number in the billions), sometimes entire governments and various ad hoc groups have been putting pressure on the contraceptive effort for practically as long as the system has been in operation. It is an ironic fact that the moment in time that effective birth control became available was also the unprecedented time of the worst population explosion ever.

The whole family planning enterprise, in fact, has, in most places, been carried out almost on the sly and on the cheap. No government has ever really gotten one hundred percent on board, for long, with the effort. Indeed, funding in recent years has been trending ever downward. The ideal system should be focused on educating, empowering and supporting women and making access to the family planning tools available in the expanding countries. No coercion should be necessary. Almost all studies show that women, when provided with sufficient information and support, do not desire large families. Desired family size is mostly culturally determined. Women gauge the norms in their society and usually go along with that as their desired family size.

This fact offers a powerful tool to use in the proposed population reduction effort. If their society and state would adopt the general idea that reducing population is necessary and a good idea, that alone would give aid and comfort to all women. Much of the contraceptive game, besides providing reasonable access to the methods, materials and information needed, lies in simple outreach, education and support. There are various well-known incentives available to be employed, too.

So, let's briefly summarize the contraceptive and family planning tools that are used. The most common contraceptive means are birth control pills (OCs) for women's use. While they have changed (and improved) over the nearly 75 years of their existence, they remain essentially the same. They comprise one or more hormones or hormone antagonists that are designed to affect ovulation, the uterus and/or the menstrual cycle. Most of the more common pill (or injection) types are formulated to subtly inhibit progesterone-estrogen cycles such that ovulation is prevented and the endometrium is rendered hostile to the implantation of any embryo. Thus, even if the egg is fertilized in the oviduct, the developing embryo (a ball of several hundred cells called the blastocyst) does not embed in the uterus, and therefore, the woman never becomes pregnant.

The two main formulations of these types, used by millions of women, are either:

(a) all month long (one-a-day), or

(b) only for about three weeks per month, which allows for a (usually highly attenuated) menstruation. This provides a "regular" menstrual cycle. Both formulations are extremely effective, quite cheap and with minimal side effects. (The side effects have been widely studied and criticized, but the fact is, these pills are about as safe and effective as virtually any medical procedure.) They have been, and are, a major tool in the arsenal of contraception and birth control and a boon to humanity.

There are dozens of these commercial OC packages, with a smorgasbord of varying hormones, dosages, timing etc., to match variations in different women. The most common one now is one called "Opill", which is (usually) just a progestin (Norgestrel). These have to be taken every day at about the same time. It has been used by untold millions of women around the world and has been a major factor in contraception for years. It is one of the few which is available Over the Counter.

There is also available a topical implant (the Patch), which works the same as the regular pill. Many of these highly effective subcutaneous implant forms have become available. These are "place it-forget it" products. Common ones are called Nexplanon or Norplant, and both give reliable contraception for up to four years.

Variations of the common OCs include a promising one, which will quite soon become widely available. This is an oral pill which is taken just once a month. This should be an excellent, cheap and easy-to-use product for worldwide use. Other variations include injectable forms. A common one is Depo-Provera, a single injection of which provides excellent protection for three months.

The most common contraceptive in many countries these days is the so-called 'abortion pill.' These pills inhibit ovulation, fertilization or implantation and are highly effective. They have to be taken every day for the best protection to prevent pregnancy, especially if it might be the outcome of unprotected sex, rape or other emergencies. It was only approved for use in the U.S. in 2000 but is already in wide use here and around the world. It has been proven safe and effective. Its main action is to stimulate the uterine expulsion of young embryos.

Its introduction and wide use have, unfortunately, been afflicted with many political and religious conflicts. It has been slow to spread around the world because of its cost and political freight. Relaxation of the stringent rules, regulations and outright proscriptions surrounding this method would be extremely helpful. In the U.S. (at least in most states), it is (was) now legal to use these pills at home. Hopefully, the huge political embroglios will be soon eased and 'normal' life can continue. Unfortunately, these are called "the abortion pill" and, as everybody seems to know, "baby-killing is a sin and murder." However, forcing a mother, often facing daunting circumstances, to carry an unwanted pregnancy should be seen as a heinous crime against humanity.

Along with the above medication abortion method, the common D and C procedure is still, of course, very commonly used in a variety of cases and is an indispensable health care and contraceptive procedure. D & C's for abortion is generally an outpatient or office procedure and very useful in a great many situations. They are very safe and effective and can be usefully done up to 20 weeks or more of pregnancy.

The above methods of chemical contraception for women virtually changed the world with their stunning simplicity and effectiveness (when used correctly). In the U.S. in 1960, about the time of the widespread introduction of the OC, the fertility rate was 3.6. By 1973, it had dropped to 1.9, or already below replacement level, and the fertility rate was still falling. These formulations are what are mainly responsible for the lowering of pregnancy rates around the world. They are also the reason why many countries are on the road to population stabilization and (hopefully) decline.

The problem with "the pill," and this goes for all means of birth control, is that a near majority of women around the world either do not know much about them, or they are inaccessible, either because of lack of knowledge, societal pressures, politics, religion, economics, or logistics. Family planning is generically the only way to rectify that problem and is central to all efforts in this area. The system is easy and cheap. It should be put to better and more widespread use. It involves, beyond the existing technology, mainly educating and empowering women.

Besides the pills, there are two other surgical methods of contraception (plus the D & C procedure). These are less desired but still sometimes needed. One is tubal ligation. By paracentesis, small holes in the skin and abdomen are made, and a stereoscopic instrument is used to locate and then tie off, or sometimes remove, the fallopian tubes. This is a one-day surgery and, of course, should be considered a permanent sterilization, so its use needs to be stringently assessed prior to surgery. An advantage where this operation is the desired means of contraception, besides being 100% effective, is that menstrual cycles and all other reproductive systems remain normal.

Other surgical sterilization approaches, which are much less used or desired, is hysterectomy or ovariectomy (or castration). When they must be used for serious health issues for the mother, for example, these, of course, cause permanent sterilization, too. For all practical purposes, they are not usually considered a major contraceptive procedure. It bears emphasising: forced sterilizations, man or woman, by any method, is NOT a contraceptive method and should not be tolerated in any population stabilization or reduction plan. It also warrants emphasis, however, that forced pregnancies, while actually quite a widespread and increasing practice in too many societies, are not to be tolerated either. Voluntary sterilizations, however, could be, and are, legitimately considered under appropriate circumstances.

Another major contraceptive method that is in wide use is also safe and effective and can be unstintingly recommended by family planners is the Intrauterine Device, IUD. These, too, have been used and in the marketplace for decades, although at much lower frequencies than the OCs. It was found in the late 1960s and early 1970s that any harmless foreign body put into the uterus of a mammal would create some subtle changes in the cycle of the endometrial membrane. These reactions would, in a high percentage of animals (and women), prevent the implantation of any embryo, thus preventing pregnancy. The main advantage of this form of contraceptive is that you can place it and forget it. An IUD can usually be implanted in a woman and left for years, providing continuous, hassle-free contraception. They can be removed at any time with no trouble, sometimes by the woman herself.

Sometimes, a useful aspect of IUDs in many places around the world is that they can be advantageous for women to have them implanted post-partum while still in the clinic, hospital or other. They can be easily configured so that there is no evidence of their being in place, so unless the woman chooses to disclose to their husband, it remains a secret. They can be quickly and easily removed at any clinic.

Over the years, the design and effectiveness of these devices have improved. They come in various sizes, shapes (e.g., coil, shield, ring and the like) and composition. Some are plastic, some copper and other combinations. Most of them now have progestins, norgestrel, and/or estrogens similar to the OC's embedded, which slowly seep out and affect the uterus in the same way as an OC to further ensure the absence of implantation. There are a large number of commercial brands with varying shapes, sizes, mixes of hormones and the like. The length of continuous reliable use is typically about 5-6 years.

Recently, another improvement in shape and composition was introduced and is already in wide use. This is the so-called Copper T IUD. It is really no different in principle from the old types, but it is more technically refined by shape, etc. It also releases copper slowly and effectively (below a toxic level), and it fits and stays within the uterus longer and more comfortably than previous models. It has the potential to be a near-ideal contraceptive for a large swath of women around the world due to its permanence, reliability, effectiveness and all-around ease of use. Its normal continued-use life extends to about 12 years and possibly more.

With all the above devices, as well as with all the other methods of contraception, it is desirable for the woman to be knowledgeable in their use. Access to medical care and advice is invaluable, too. Obviously, this condition is often unable to be met but hopefully will be largely overcome in the future. In the meantime, information from family planners to both get the word out initially and provide the means for contraception is essential. It is the backbone of the whole Velvet Touch prescription for a forever population plan. At any event, no matter what the social and medical deficiencies faced by women in many areas of the world, if contraceptives are made available, with the consequent prevention of multiple unwanted and dangerous pregnancies, it would provide an enormous immediate good to millions of families of the world, as well as the world family itself.

Several other methods of contraception, too, can be added to the list, such as diaphragms, spermicides, douches, condoms, etc. Condoms, of course, have been a useful tool for a long time but have obvious shortcomings for reliable sole use. Their continued use, however, is still much encouraged, partly too for their efficacy in preventing the spread of diseases.

A contraceptive for men has long been a holy grail of much research. Currently, the only highly effective and widespread method has been sterilization or vasectomy. A similar approach of this type for men, called RISUG (Reversible Inhibition of Sperm Under Guidance), will be available soon. In this 15-minute procedure, an inert polymer like styrene, plus various other chemicals, is injected into the vas deferens. The chemicals inhibit or kill sperm "electrolytically". In advanced clinical trials, it appears that these could provide protection for years, and it is reversible.

The last contraceptive procedure to be considered here is the oldest. This is the old rhythm method. It is still a useful method in some situations. If nothing else is available, it is much better than nothing. When practiced knowledgeably and consistently, it has surprisingly good efficacy. Even if nothing else were to be used in a large population, it would definitely result in fewer pregnancies than naturally. It has one distinct advantage over all other methods of contraception: it is very cheap. It costs virtually nothing. The price for the woman, however, in terms of time and attention and knowledge and support etc., is quite high; too high, in fact, to rely on for any kind of sole, wide recommendation.

This, then, is the contraceptive tool kit of the world— the Velvet Touch. This is all we need. We don't need any new scientific breakthroughs or improvements. We don't really need any new materiel, or equipment or, techniques, or protocols, or personnel types or procedures. The tools have worked already to bring many countries of the world into near or below replacement rates. Without their use, the population would still be growing even faster everywhere. From now on, essentially, all we need to do with these tools is keep up the good work. And expand their use and scope, especially into areas of the world still without them or the wherewithal to use them. These tools and this work will serve admirably for a thousand years. With their help, any or all countries will be able to slowly and steadily reduce their population generation after generation until a hopefully sustainable, quality living number is reached. (We might note that populations could be readily increased, too, by basically similar education methods. This should allay any fears of an empty planet!)

So, this is the grand plan. Simplicity itself, it would seem. We can envision a simple (or outrageous, depending on your outlook) country-by-country program that aims solely to painlessly keep pressure to use these tools to achieve and maintain slightly negative population growth, i.e., degrowth. In the U.S. and all the other countries with TFRs below 2.1, for instance, there is no need to change infrastructure or methods that currently oversee or administer the technical aspects of this subterranean birth control methodology. Just ramp it up and keep on doing it. If this program is part of the social and governmental approved list, that would help immensely. It could bring any population (even Nigeria, say) up to speed quite rapidly.

All of these methods have been tested in various countries and have been confirmed as safe and effective. All experts agree with Cafaro & Crist. (2012), who states: "Wherever modern family planning is made available and barriers to access lifted, women and their partners almost universally choose to have fewer children."

As we stated before, for nearly half the world's nations, there is little need to do anything greatly different in the way of carrying out a population reduction program—they are already doing it, even if often-times unknowingly and without any overt program. Their people do, however, have to come to understand what they are doing and, most importantly, also develop a socio-economic and political system that does not depend on growth and an always increasing utilization of nature's resources. The other half of nations, though, have to develop and implement a plan to reduce their TFR to 2.1 or below. There cannot be, in any nation, steady increases in anything, including people, tin, iron, copper, water, and anything else we all use for life. The Earth cannot support it, pure and simple.

The whole plan, then, all over the world, is really just a scale-up of the present 'family planning' platform that is now in (anemic) use in multiple dozens of countries. We've said in many places that the task should be easy. The plan itself is plain and easy. But deciding to put it into practice at the scale proposed is not easy. Doing this will be the hardest single program that the world has ever done. Even getting it started will be the hardest thing most countries have done or ever will do.

Then, the hard part! Dealing with the fallout of the results of implementing the plan. As we have noted, the tentative little steps taken by some countries already that have resulted in, actually, the desired outcome have created some little panic already. The economic and political zeitgeist have been exposed to be unable to handle the new social dynamics that lowering (even stabilizing) the population causes. The knee-jerk solution, so far, seems to be— stop the plan!

Eventually, no doubt, from time to time and in one place or another, some attention will have to be given to "Cheaters". Certain groups, religions, counties or countries, etc., may (probably will!) try to conduct Reproductive Warfare in order to gain influence, power, etc. That is, they may encourage higher fertility rates to enable them to simply outnumber their supposed enemies. Some forms of sanctions could undoubtedly be instituted, but hopefully, these outbreaks will not be extensive or numerous. For rogue nations that create global havoc, various forms of persuasion and/or sanctions could be envisaged.

What is necessary now, however, is simple: decade after decade, generation after generation, keep doing the same thing all over the world (perhaps needing a little rate-adjusting from one decade or century and in one country or another). The endpoint desired is when that country's population fits into its total ecosystem and biosphere. Until it reaches at or below generous, natural carrying capacity. It is important to emphasize this last aspect of carrying capacity, which has been covered in this book. We do not want just to minimally balance the resources of the natural world to the numbers and activities of the 'good' species (i.e., us), which is the usual measure of carrying capacity. No, while we are at it, the desire, expressed so well and in so many ways by so many recent authors (e.g., Korten,1999 and others referenced here), is to "live a quality life, in sync with the natural world". A society could, rather readily, one would think, be invented that is built around community, for example. One that has self-interest served rationally with some form of cooperative action. And each country should essentially be able to feed, house and maintain itself, largely without massive foreign aid.

At any rate, we need to find ways to continue to invent ourselves into a better, feasible world order that we can all live with.

Chapter 10
To Build a Better World

So, to this point, we have taken a broad "tour" of the world to see what the effects of teeming humankind have been on the physical and social ecosystems. We, as well as many others, have concluded that the physical underpinnings of our civilization also is becoming completely unraveled. Rampant biological extinctions, dangerous warming, and impending massive shortages of food, energy, soil and water are readily visible hazards to continuing civilization as usual—our existence even. We, therefore, along with so many others, conclude that our best (indeed, our only) hope for saving ourselves from a dismal and likely fatal fate over the next century is to build a better world. Unfortunately, you, dear Reader, as well as most of the world's people, probably don't believe it either. It appears that the whole world apparently needs another 50 years or more of increasingly painful human and ecological disasters happening in real-time to see the obvious.

However, be assured the world really is at a crossroads. We have been rather ignorant in a lot of ways. Blindly and blissfully, we have followed our noses and occasionally our finest instincts even. Astoundingly, we have remade the world and done so many incredible things. We have come very far. But we now must know that ignorance of some of our flaws and mistakes is no excuse. Some new ways of doing things must be invented. There is a warning bell ringing loud and clear!

Never before in human history was there any need to do such a global act as we are asked to do: reduce, not grow, the population. And all this to do while developing a world not dependent on growth. The test now is to not only try to improve the quality of humanity and civilization but even how to save the whole thing. It will require nothing short of heroic action. This is a pivotal moment, the most dangerous time in human history. No greater crisis has befallen the world since the last big asteroid struck. We need to get started revising and re-thinking so many nation's policies and ways of acting and thinking by all peoples everywhere. We need to generate new ideas and means to de-grow in some areas and keep growing in others. We have become accustomed (maybe anesthetized) to our natural rise in health, wealth and goodies. We just "knew" that this cornucopia would keep on rolling along. No need to worry or change anything very much.

But now, we do need another kind of historic revolution, the Third Revolution (or Fourth?). We, the last several generations, have already delivered to history at least two previous fantastic revolutions, e.g., the mighty Industrial Revolution and now the second, gobsmacking, Digital, or Information Revolution. They have been good to us. Barack Obama said, with some justification, that this is the best time in all of Earth's history to be born. We might very well be entitled to pat ourselves on the back. Our super-brain, super-species, has indeed produced many awesome and laudable things. Things like art, music, philosophies, superb structures, tools, education, etc. etc. We have produced Aristotle, Newton, Darwin, Shakespeare, Mozart and thousands of pathbreakers in our heads to give us amazing good and human things. We have also, obviously, done and are doing so many stupid and rotten things. If we don't shape up real soon, we will create our own certain doom with our own hoggish hands.

So, we are now heading down a broken track. Mindless growth and money mania no longer improves humanity's well-being or the common good. The last madly-growing- and-prospering centuries haven't brought us just all good things. There have been huge, often hidden, costs and damages, both to our Earth and to some parts of human life. For practically every one over a certain age, the near incessant adjustments to the unsettling light speed of change in our lives, our surroundings, our environment is a constant nagging fact of life. However, the future looks nowhere near as rosy as the recent past. In fact, if we don't wake up to the new reality already settling in, we won't have a future. As Leakey (1996) said, "*Homo sapiens* might not only be the agent of the sixth extinction but also risks being one of its victims." Chris Hedges (2015) said, "There is a familiar checklist for extinctions. We are ticking off every item on the list." But just exactly, how in the hell are we supposed to do something as endless and hopeless and outright cockeyed as this, you ask? You are threatening to downsize civilization itself! This is outrageous! This could throw every nation into economic chaos and disaster'.

Indeed! It probably will for a while. We *are* saying that civilization needs to be downsized and rebuilt. But mainly in its total size and some of its accrued unsavory aspects (but hopefully not in its totality, a la' Jensen, 2006). But we must rebuild civilization and change, in meaningful ways, how contemporary (and the following) civilized human beings operate in the social and Natural sphere.

You think we have a choice in this. But actually, we don't. We are, in fact, currently acting as in a fairyland where people eat themselves out of houses and homes. Either we resolve to pare down and allow nature to survive whole, and ourselves again be a part of a functional, natural world, or we will suffer such angst for a miserable couple centuries as the Four Horseman could barely dream—if we survive at all. It would be nice if we could come back in a century or so and either say "Told Ya" or offer profuse congratulations on advancing acculturation and see a more satisfied and happy race of men.

A. Where are we, Again, & How did we get Here?

It really should be obvious that we have arrived at a critical point in the history of both Man and the world, a point at which our old ways will no longer serve either Man or Nature. How did we arrive at this point? We'd answer, "Exactly where all our history has told us, and the world, to go." But now, must we basically begin undoing so much of what we thought was the finest and best humans could do? No, not undoing; just re-adjusting a whole lot of things. We, and our parents, grandparents, great, and great great grandparents, worked so hard to grow, 'tame the land, be fruitful and multiply' and build great things. They (we) served the god of Progress and grew large and wealthy because of it. The god of unbridled free market capitalism based on the growth of everything was mainly the primary and proximal engine (unknowingly and unwittingly). This, however, was made possible only by low but growing numbers of people (and earth's abundant, readily available energy and larder) through most of that period.

Hopefully, we will soon embark, using our cultivated Mento-Sphere minds, on Enlightenment II to begin the historical journey to solve not only our own ecosystem's ills but improved ways of living sustainably in and with Nature and ourselves ever after. We will continue the spirit of the Enlightenment, keeping some of the good and ever-improving what is not so good for the next 1000 years. The past couple of centuries of mindless growth and progress, which led us to unwittingly become the mind of the earth, is, hopefully, not the end of what our minds can do. As we have been attempting to show here, it is the mind of man that must be put to work in service of minding the whole earth, not being the witless plunderer, conquerer and crowned King of it. Taming Nature seems to have been in the stomachs of Man for at least several millennia.

It will take decades and decades to reduce Man's numbers to a safe level, as we posed earlier, to the eighteenth-century (or earlier) level. After all, it took centuries to get us to here. At the beginning of our country, it was a big, empty, green, bountiful and (almost) clean slate, and (the invading) people thought they could do anything to the land, just about as they liked, with no heed to the possibility of despoiling or depleting it. Their progeny, our immediate ancestors, just like ourselves now, followed the same mantra. Any unoccupied or unused or undeveloped land was (is) considered as going to waste, and worse, an indication of sloth. It was all built on the notion of growth, of progress, of new horizons—and greased with "things" and money. Subduing nature (clearing the wilderness) to bend it all to our own expanding social, economic and psychological horizons was the near-universal unspoken mantra. Everything was limitless.

Grandin (2019) traces this all-encompassing American frontier myth all the way to the End of the Myth. Movement West into this boundless space said Jefferson, "Wasn't just a fruit of freedom; it was the source." Madison said, "Expansion didn't threaten the common good; it _was_ the common good." Thus, Grandin says, "Freedom depended on expansion, and expansion becomes the answer to every question." The notions of this paramount frontier myth embodied unfettered capitalism, with its power and limitless possibilities. This notion was early on put into moral terms and was (is) applied to not only land expansion but also markets, wealth and militarism. It was also based on the surety of no limits to wealth and goods and power. All the way to (at least) Ronald Reagan, who gave it a modern, giant shove. "There are no limits to growth," he said. "Nothing is impossible." Now, however, Grandin notes that the great and triumphant frontier and American myth is finally ending (and also uncovered as being not really truc to boot.)

Unfortunately, a great many people still essentially feel that warm fuzzy feeling of America the Great; we seem to all want to go, or stay, with "that". We still live with a lot of our 19th-century cultural minds. We know better now, though, don't we? That we are making some very bad mistakes and seriously degrading our own and all of nature's territories?

In some ways, we might excuse ourselves in our wanton treatment of our physical world by "We didn't know better." Because it has worked. It has been an astounding success. Until now. As EO Wilson said, "With scythe and fire guided more by ignorance than by reason, we changed everything." (Of course, all this was not restricted to the U.S. It is a worldwide phenomenon.)

We think that it is necessary to strike at the root cause of the problem for a full and permanent fix, i.e., lower the population numbers. Then the problems will rather quickly start to get better and eventually resolved, without having to apply gigantic, damaging, expensive and never-ending geo-engineering, Nature-chewing (and eventually hopeless) ameliorative projects.

Pankaj Mishra (2017) has brilliantly traced our Western as well as Eastern Civilizations from Buddha to Rousseau and everything in between (including Trump!). In America especially, right from the beginning, as the historian Greg Grandin (2017) also emphasizes, it was an amalgamation of various triumphalistic efforts toward endless expansion of everything. Especially endless economic expansion and private wealth accumulation. It was a mash-up of a wild assortment of ideas and feelings, leading, Mishra says, to our current 'Age of Anger'. Grievance and anger (the 'ressentiment' of Rousseau) drives to a considerable extent, our current political direction and "reflects a severely diminished respect for the political process itself."

Mishra also points out that up to now, "There, plainly, has been no deep logic to the unfolding of time." [And we might add, or to history.]

So now we are looking at a double whammy—not only are we waking up to a Natural armageddon, but also a coming political and social nightmare. This is by far the biggest challenge that Mankind has ever faced or consciously and intellectually recognized. Unfortunately, as Stewart Udall et al. already back in 1974 noted, "All the available evidence says we have already passed a point of no return, and tragic human convulsions are at hand."

Now, we need to consciously and intellectually attempt large-scale reconstruction of our human societies and, as a result, of Nature. The effort may not be as much fun as the last two hundred years for

Americans in producing wealth and goodies and easy living without hardly even giving it a thought. But it will be worse if people do not take the steps to come up with some new rules.

Thus, Mankind, as he has culturally evolved, has muddled along as best he could. Looks like we now must just keep muddling along, and hopefully, eventually, we might come up with excellent results. We are stuck now and will need to just keep plugging away, trying out new forms of social and governance organization. Fortunately, we have accrued some additional very powerful and interesting knowledge and tools that we could use. Some of the things our history has produced in the last 10,000 topsy-turvy years, like democracy, equality, liberty, science, governing styles, etc, we will surely want to use and improve on. Heinberg (2011) says, "If civilization fails, it won't be for lack of good ideas."

B. Changes in Society & Government Needed

We do, indeed, need to do some downsizing and make many things in our world smaller. The ecosystem can't handle it as it is. And downsizing needn't actually be impossible to do——or to live with once it's done. If populations fall, downsizing will be more or less automatic, and everything can begin to self-correct in many little things. Fewer and smaller cities, smaller farms and industrial parks, fewer cars, parking lots, land developments, fences, malls etc, etc.

It is true, for people as a whole in a few countries (increasingly fewer countries, however), things seem still generally going in the right direction materially. For Nature, the reverse is true. The system is especially not working for the biosphere and environment (nor for a large part of the people in most countries either). Even if our own (highly contrived) environment doesn't suffer major collapses, it seems unlikely, without some changes in our societal and governmental structures, that we could avoid serious rolling disasters (and not excluding horrendous human potboilers of a familiar kind).

We have pointed out a few of these rather localized, collapsed and near-collapsed ecosystems earlier. When they happen in our own little piece of the earth, they are catastrophes. And when they are not recognized, they will keep getting worse. Even if gigantic human industrial, engineering and scientific mightiness would be mobilized, for some of our galloping water woes, for example, it won't help much. In fact, in many of these cases, adding more of the same kind of medicine that led to the crisis would make it even worse.

We need to develop systems involving planning, ideas, theories, procedures, and nuts and bolts administrative rules and regulations that will work to preserve our natural environment while also providing satisfaction and well-being ("prosperity") for all. Derrick Jensen aptly said, "The industrial economy is in competition with human and non-human freedoms, and in fact with human and non-human life." Similarly, McKibben's new book, ("Falter"), delivers the loudest clarion call for a new beginning that you can get.

It is time to try approaches aimed, this time, also at soft targets, like economics, government, social structures, social evolution and the like. We now need some universal, true guidelines for the new governments and societies that we will need for the next centuries. Not just for financial rewards, as mostly was the case in the past, but also for making possible a healthful, rather easy, happy, fulfilling life for all. Crary and Paul Mason (2015) are among many who similarly call for this kind of post-capitalism society and civilization. Population reduction, however, is still the first key step in a rational, conscious process for us humans to design a safer, better earth.

Tim Jackson (2017), Lester Brown, Jeffrey Sachs (2005), Richard Heinberg (2011, 2021), David Korten (1999), Herman Daly and colleagues, Peter Whybrow, plus many others around the world have already put out lots of great ideas for theory and practice to accomplish change. Peter Barnes (2006), an entrepreneur, has published, in great detail, plans for many new programs. He calls, for example, for Capitalism 3.0. Capitalism 1.0 was the first real economic system of Adam Smith and John Locke essentially. Goods and capital were scarce, and corporations were at the nascent stage.

Capitalism 2.0 is what we are in now and has been for a couple of centuries. That has morphed mostly into naked, unbridled capitalism fueled by growth, goods and greed. Goods and capital are not scarce, but consumers with enough money are. That is why we keep needing more people (Kelly, 2001; Jackson, 2017). The nature of Capitalism 2.0 is mostly to maximize profits (almost always by minimizing costs) to maximize returns (wealth) to shareholders. Large corporations and banks, however, ended up increasingly being the holders and maintainers of almost all the capital and wealth. Thus, by nature, it returns most of the profits to already existing wealth holders. Between 1983 and 1998, Barnes says, more than half of this wealth went to the top 1 percent. Further, the top 5 percent own 95% of property wealth. With wealth goes power and, thus, the urge to set the rules. This is what this kind of capitalism does; this is what it will continue to do unless new tweaks are added. It also totally ignores the actual costs to the environment.

Crist, too, (2012) boldly says that human supremacy (arrogant ignorance) fuels the top-down conceptualization of Nature as a Resource base and domain for our own use. She insists, though, "that Earth is not just a resource, or property to serve a populous master race."

What we did in the past few centuries produced much good. But the physical progress we have made lately was mostly geared toward limitless wealth and comfort, which is not inherently bad but loses something. Now, we are running out of room and materials for relying mostly on physical and monetary systems to grow and drive ordinary life. And, most damning, we are crowding Nature out of the room and profligately using up her stored resources. We can't keep on keeping on just the same. But, as we advocated, we might be able to keep much of the basic progress we've made and keep on moving on. But now, in our own little piece of snatched history, we must also pay much more attention to the social side of humanity as well as the physical side of all our earthly neighbors and homes.

What we're trying to say here is that our problem is biological, too, not just political, social and economic. That is, attitudes, personality and sociality themselves, in all their myriad and confusing aspects, are the result of the day-to-day workings of the brain. Much of that "working" comprises broad patterns of behavior and thinking that have been baked into our genes and basic human brain architecture through cultural and biological evolution. These have been endlessly modified over time and from over our past civilized history. But these kinds of changes that have been made are operational and can be modified and passed on generation by generation through education and society in general.

We are now starting to faintly understand how, over the thousands of millennia of human as well as animal history, how those "lumps of protoplasm" we now call "brains" actually can work to produce its gobsmacking product. This product, of course, is the immaterial things it produces, like thoughts, ideas, imagination, feelings, art, religion, science, understanding, etc. etc. This is where the ancient hunks of rock, water, and air eventually came together to produce the first living thing that leads to our ('aware') brains and concomitantly on our present planet. How could these have morphed into our mammalian brain, and these, in turn, eventually result in the astounding arrays of behaviours? How could ours have then come to, even more inexplicably, produce our ineffable range of activities, like philosophy, mathematics, computer, logic, electricity, mega construction structures, and contemplation of the entire universe and our own origin and fate? Albert Einstein, way back over a century ago, expressed this most astounding nature of our human brain. He plainly said the incredible truth: that "pure thought" can now grasp the totality of reality itself. There is nothing in the Universe more significant or ineffably magnificent than that.

We may never know about all these things. But we will soon have enough information to turn at least some of this knowledge of how our brain does some of this magic stuff to actually use it to our own advantage (as well as the rest of the planet, too). We will, of course, also have to reckon with the clear fact that we are still made of the same old clay. As Ruth Marcus (2019) and others say, "Some assembly is required for people's brains to enable them to invent and operate a higher human plan for a society or state".

The entire educational apparatus is the main mechanism that most societies use to transmit these psychological and cultural 'mindsets' to the next generations. Recent insights and methods from the neurosciences now offer some more advanced and powerful tools that society could profitably use to facilitate the process and extend the reach to almost every individual. Science, of course, is what humanity has lately relied on to produce much of our current physical progress— and some of our problems.

Artificial Intelligence is also making rapid improvements in technology. It is considered quite feasible, if not desirable by some, to advance this field to the point that robots and human-machine interfaces could become quite useful in many ways. However, this kind of thing could also be imagined as helping to lead to sci-fi mayhem or even the end of humanity itself. As we work on all these new tasks, we should also try a few little tweaks to human nature to mold societies based more on the sunnier side of nature, like love and laughter, compassion, kindness, sociality and the like, not pander to the ego, dark, homophobic, or domineering side. Societies and communities have to be crafted, not slapped together through the vagaries of tribal or national strife. If we can comprehend the entire universe and many of its workings, we should be able to live long and well if we would put this to work.

C. More Suggestions for What to Do

In the meantime, while we attempt to adjust the numbers of our population to match the biosphere demands, we must also continue to struggle with stopping and repairing the damage we have already caused to the earth and ecosystems. The tools to achieve population reduction are rather simple and near at hand. Some of the tools for starting to deal with the repair job to our atmosphere and biosphere are also if truth be told, near at hand. Some people are already at work at the task, in fact. Ideas and methods for reducing the amount of carbon dioxide entering the atmosphere, for example, are beginning. If the population is lowered, some of these current approaches will be increasingly helpful.

But the fix for the physical ills of the planet is not merely Technology. "No technology can maintain the way of life, so many have gotten used to" (Heinberg, 2010). For our suggested remedy, nothing really new or exotic in the way of scientific or engineering tools or knowledge is needed either. As the Ehrlichs, EO Wilson (1997), Smail (2002), and dozens of other leaders, groups and organizations have been preaching, getting the will to even begin the urgent and daunting task will be the hardest of all. Placing man back into his necessary role in the biosphere is only feebly beginning and may be the hardest part of all.

In the beginning, as noted many times, the immediate problem will be economics. The mantra of growth, growth, growth must cease. It has worked great for centuries to make the world as it is now. It is actually all that everybody living now (or their great, great grandparents) has known. And, until recently, we haven't felt any good reasons to worry. But it will not work in the future. For sure, our past and present economic-free-for-all is the antithesis of a Nature (or even a decent civic) ethic. Our age might be the Anthropocene, and our times the Digito-Sphere, but our tools and progress in economics, government and advanced societies are still basically Stone Age. Diane Ackerman (2014) asks what sort of world do we wish to live in and how do we use that Anthropocene knowledge to design new, human-made shape(s), like relating to our spouses, children, friends, coworkers, etc?

When the population actually does start to go negative, and people can see some problems arising, and even immigration stalls and stops, then we will hear the mantral hue and cry from the usual suspects even worse than presently. But we must persevere and invent an economic and social system that is not broken by it. The mantra of perpetual growth and the old and present obsession with the fuzzy notion of *Progress trumps alles* must be mostly abandoned. All they really seem to boil down to is *More Money for Me.* Materialism need not be synonymous with greed (though it may possibly rhyme). Jonathon Crary, who posted many good suggestions for how to create better societies with emphasis on human compassion, personal well-being and community, says, "The struggle for an equitable society calls for the creation of social and personal and local arrangements and abandons the dominance of the market and money." Gar Alpervitz's book also contains a toolbox of practical things that "might one day help realize a U.S. politics around Plurality Commonwealth."

Everyone should be reminded again that growth in most things would cease almost immediately after population growth simply stops, let alone decreases. We will already be, de facto, in a steady-state economy at that point. That has been amply demonstrated already by the angst that this situation has produced in several of the countries we have mentioned. The big engine, more growth and more jobs (the big "J" and "G" for politicians and economists), which seems to be the driving force these days for all ordinary life in all countries, will sputter and wane.

The economist Kenneth Boulding stated 50 years ago. "Only madmen and economists believe in perpetual growth"

Many other people, including Barnes, Bregman (2017), and Daly (1977), have pointed out ways to achieve this new economy and social contract, one with prosperity without growth. What those new systems will actually wind up looking like, we don't yet exactly know. They must be carefully developed. There will undoubtedly have to be some kind of circular economic system that is based on the growth of nothing except ideas, well-being, health, moral reasoning, art, equality, and respect for all living things and the environment. And all actors will have to be held accountable by their society. Education, low-impact technology, travel, community activities and such will be some of the mediums for determining well-being.

The road ahead, no matter what route mankind takes in the next generations, will be difficult; the most difficult task ever faced by the historical human species. Up until now, we have been essentially running on automatic pilot. That is, largely heedlessly, mindlessly, swimming along on blind social waves, only slightly checked by cultural evolution advances. Work, play, get ahead, get rich, become kings of their area, grow, grow, grow, advance, change, do whatever you feel it takes (e.g., rake in ever more money per square minute is the current mode). Economic expansion leads to prosperity and higher income and is automatically equivalent to a higher quality life. Business as usual, we assumed, would lead us to greater and greater heights of wealth. That was our unspoken, unthinking motivation. Grow! Progress! Noble work! God's work. We had an open book to operate in, as the world was still largely unpopulated and appeared to be open and inviting to be bent to our every whim.

But our progress has come with a cost, a huge, unbearable cost in our very living milieu (both the natural and social worlds). A cost that we still haven't clearly seen nor heeded. We can't do it quite like this anymore. Sorry. And we might eventually find that consumption and income does not equal real prosperity after all. But what we must do, is eschew doing business as usual and start to build a new, permanent kind of human lifestyle that will begin to heal the damage to the planet that we've already afflicted so all people can live more happily within our biospheric means.

And yes, you are right. We are asking you to undertake the largest, most disruptive, most consequential acts by far in the whole history of civilization. And you are right, too, in how difficult and painful that will be. However, you will have the satisfaction of knowing that you are doing a noble and necessary job. This heroic accomplishment, part of which will be to cede back much of control to nature, would carry the following generations through. A factor to be kept in mind (and our spirits up) is that if the population is reduced by one-half or two-thirds or more, there will be less traffic, and "stuff", cars, airplanes, trucks, garages, telephones, gizmos, etc. etc. to manufacture, clean and service. By itself, this represents a huge boon for the energy, water and soil environmental benefits that would begin to be seen fairly soon. Fewer garbage pits and roads to build and services will be needed too. Fewer housing settlements and factories, etc., would be around. This lowering of demand should also be a great aid as new, people-friendly social structures are put into play as the situations arise.

Some will say the government can't force people to reduce the population. No, they really can't. But an enlightened and more rational populace can demand accountable and reasonable government. And in that sense, Yes, yes you can; society has every right to, collectively and willfully provide for the common good ("ye olde pursuit of happiness"). However, collectively and willingly, society must provide for environmental and human well-being as well as money, bread and home. And, of course, trying to force ourselves to live apart from the 'environment' is our original sin.

Continued and expanded cultural evolution in economics, civics and social structure is imperative. No more can the environment be simply a pawn in the frantic drive to make money. Pure, unregulated or poorly regulated capitalism as the driving force of society, where the drive is primarily for making more and more money for oneself, is a recipe for the worst-case scenario. The moral reasoning necessary to regard the natural world, our only environment, as an integral part of a sane and happier society is a major part of a needed psychological evolution. Regarding it as a commodity to use by any person or group capable of grabbing as large a slice as they can as fast as they can to produce monetary wealth for themselves only is a past and current (and ultimately fatal) disease.

"The prevailing vision of prosperity as a material Paradise has come unraveled."
–Tim Jackson

This is where we are at.

D. Last Words

In this book, we've tried to make the case, using accumulated technical and scientific, as well as naturalistic and humanistic data, that we humans have created a sadly weakened and imbalanced, dicey world. We included some speculations and thoughts about the health and safety (or lack thereof) of our current human situation and ways of thinking. And we've argued for some fundamental shifts in those ways of looking at nature and our own social contracts. The message is simple: Man has grown too many for this world. On the present course, we are staring into the maw of a ghastly future.

Our message is also one of hope. Actually, our approach could be the last best hope for a rather easy, though long-range, way out of many of our created messes in the biosphere. It does not appear to many of us that there is a credible way of solving the disastrous environmental upheavals coming our way by way of either standard social or scientific or technological bandaids. Keeping on doing what we are doing today is the surest blueprint for disaster. It seems that we must retrace the impact of the past few centuries. That is, we must reduce the causative agents that created the intractable problems and ecological mayhem.

By gradually reducing the number of people who are inflicting the damage, the damages will automatically be correspondingly decreasing. Further, few new increases in the damages will be occurring, so the currently envisaged amelioration activities for CO_2 and climate change, for example, can also be proceeding apace and more quickly avert some of the dire consequences of previous injury. At that point, sequestration of CO_2, for example, could indeed greatly slow down the CO_2 (or methane too) going into the atmosphere and (hopefully) stop the temp from rising any further. Air pollution then would soon also begin to rapidly decline, resulting in improvements in our health and energy domain. Ecosystems (Nature), too, could begin to operate more normally and aid in restoring balance to agriculture, soil, water and other aspects of man's activities.

Now is the time to make some fundamental changes in the ways that societies wish to proceed. The old way of financializing everything, markets and profit-only aspirations, with all the weight of the financial, industrial and governmental inertia behind it, is getting too heavy for Nature and people too. Our old order will not fit entirely into the new human situation. It is time for a new movement to bury Hoover and the current (old) order; maybe try a new model, like little 'c' capitalism 3.0? Lester Brown called a long time ago (1974) for a "Workable New World Order," as did Gar Alpervitz (2011) for a "Pluralistic Commonwealth." (Along this line, it seems kinda funny, but it appears that there has never been any really fundamental theory of government that we could follow).

We've got a lot to work to do but also a lot to work with. It all essentially depends on how we arrange our goals as human beings and how we build our social structures in alliance with more noble, sustainable human qualities of life. We have ways and means sufficient to start the evolution. But, it is for you, especially the young people of today and the next coming generations, to repair the damage that our activity, our thinking and society have caused to the biosphere. It will require, actually, for you to partly remake civilization. It will be a new, better civilization. People could live comfortably with some of the modern luxuries and materials we have, but many things will have to be different.

There is more than enough expertise around to invent a new order. The time has come to end the hegemony of capital and markets and corporations and ruthless competition over government and society and Nature and the commons. Making money from money is now the high road to success, i.e., success flows from money. Noam Chomsky has called all this the "neoliberal destruction of the social contract." The really good life is blatantly incompatible with 15, or 8 or even 4 billion people. Painful though it might be, the cost of reform will be far less than the price tag of the status quo. In any case, you cannot buy your way, nor harangue, nor science, or BS your way out of this one.

But we recognize that there are those who would say that our proposals are totally out in left field, from la-la land. Way too extreme and radical and wacky. What you are suggesting is totally impossible and crazy and unreasonable. Get real! You guys are unrealistic, hopeless zombies and dreamers. You don't know how 'real' life is, with business, and industry and commerce and finance. What it takes to keep the wheels of business going, to keep people working, and clothed and fed and educated and all the rest. You can't just snap your fingers and disrupt the very essence of everyday life at the drop of a hat"!

To that, we would say: "Don't call us the hopeless pie-eyed dreamers from high-flying flights of fantasyland. You, you are the unrealists, lazy magic-thinkers. We are, actually, the non-dreamers, the hard-headed, clear-eyed forward, not backward lookers. We see the course taken by the blinded, short-sighted and greedy owners and keepers of wealth and power and tradition who, for the past centuries, have heedlessly plundered and despoiled the earth with the too-much of Cornucopian pipe dreams." In the [slightly modified] words of Shakespeare, "Behold yon [earth], sickli'ed o'er with the pale cast of [Cornucopian] thought."

The old, now thoroughly entrenched (sloppy) ideologies, and ways, means and habits of the old generations are becoming now a blight on Nature and a direct threat to the well-being, even the lives, of all people. (What part of Bartlett's axiom do you believe that you can BS or dance and weave your way around by some sort of economic or scientific wizardry? Do you really think the things on earth have no limits? "Limits, hell! Grow the limits").

It is, then, high time that the new, the young generations, with a perspective and philosophy different from the old devotion to growth and power and markets, must begin to take charge. For the good of the majority of people and for the Earth. They, the present rote, back-looking owners and operators of the world, must slow down and let the young generations get into the act of fashioning, eventually, a stable, realistically and reasonably grounded society, philosophy, nature and operating system. Smug, witless and blind pursuit of money, growth, and tampering and gouging natural systems is destroying the web of life from which we ourselves arose.

Sadly, it seems the current owners and operators evidently are not capable of doing the job. We need to start replacing them with the real A-team; with fact-grounded ideas and plans. The old guard has gained power mainly by parroting platitudes plastered over porous tribal beliefs. The old grind of homespun, down-home, common sense good ol' boy know-how that somehow seems universally and infallibly true to 'just know' the proper natural order of things. "Common sense" you know. As John Muir put it over a century ago "Sound, sensible, irrefragable ignorance." This must be replaced by something based more on facts, reason, reality, empathy, community and truth-seeking. Otherwise, revenge of the real world is coming soon.

The argument will be made, of course, that population reduction is not really the answer. The argument is already being made that since population growth is already slowing and may well become static in a century or less, that the problem may solve itself (e.g., George Friedman, 2004; Bricker & Ibbitson, 2007). Then, we could apply all our scientific and other technical fixes to the residual environmental and other problems, and then things can proceed, much like today.

Our answer, again, of course, is: "baloney. Why?

Why are you OK with a stable population of 9 or 10 billion but not with one of 1 billion? Even if there were 9 billion, or 7, or 3, you would then still be, necessarily, in a steady state economy, and nothing really would have been accomplished, except you have dwindling resources left and a still broiling population and an increasingly damaged world."

Kolandreowiscz (2012) puts it nicely when he says, "Limiting the human population would not guarantee success, but not doing so will guarantee failure."

The world pandemic and political and social crises should stimulate and create new ways of thinking and proceeding in the public sphere going forward, too. Obviously, the resulting current social and economic crises will greatly affect the ongoing natural, social, economic and political actions for the next decades in any case. These crises alone will sooner or later create huge changes in many of the core tenets and structures of all countries in the world, perhaps the most profound ones here in the U.S.

In fact, though, the fracturing offers an unparalleled opportunity to build a better, stronger, sustainable socio-economic system. The old wormy screed of the hegemony of money and markets and the econo-political theory we've been under the spell of for so long seems now shallow and of tarnished value. Our incredibly equipped and talented species could, one would hope, develop better, more community-oriented, less corrupt and inequitable operating systems. For one thing, that much-abused word 'democracy' needs to be polished up, put front and center again and re-lustered.

As recent events have clearly shown, the creeping environmental damage, etc., has laid bare the social and economic and governmental inadequacies. Our recent history in inflicting so much damage on the natural world and environment, as well as increasingly the human world, has become intolerable. Willful ignoring of and uncaring contempt for the consequences has become inexcusable. The hollowness of the human conquest of the world is becoming clearer.

We solemnly have to say, today it is unethical to act or pretend that anything close to the current number of humans can be sustained or be satisfactory to us and the animals and plants we have evolved with. We're now in uncharted waters. Since it clearly appears that there is a need to create many new systems, plans and programs for all aspects of the economy, it would seem very wise to make a new tomorrow, not just play a rerun. A return to normal is not an option; normal is not working anymore.

Normal is broken. Perhaps, though, a breeze could be beginning to blow (the coming Winds of Change——Revenge of Gen X?). The seeds of demand for fundamental changes are possibly faintly planting, so the new tomorrow could be much better. Prospering and well-being without growth would be a great gift to future generations and to our natural world. The new generations will be seen as the heroes in the future, the ones who created the crucial breakout of a sustainable civilization.

The new direction for the earth that they will take has a much better chance of making the earth habitable for the next 1000 years. Most of the current owners, apparently, are hopeless and essentially clueless about the future. If the old guard owners won't, or can't do the job, as is already proving to be the case, a new direction and ways of doing business needs to be instituted by the new generations. The old guard should step aside to save the planet from further, drastic damage. Start towards a new path of human cultural evolution based on the good and the bad accomplishments with better, higher and moral principles of society. Crary (2020) said it best "……..amid the burning and pillaging of our lifeworld, there is little time left to meet up with a future of new ways of living on earth and with each other.

Population reduction, we say again, "is the only long-term way to enable us to dig us out of our pickle. Start doing this simple thing—or else!"

It is time to hit the Reset button while the time is ripe and do two things immediately:

(1) Apply the Velvet Touch worldwide.

(2) End Growth as The Source of everything and start to build a steady state socio-politico-economic operating system for sustainable, improved societies, civilizations, and the 'lifo-sphere' worldwide.

In conclusion, the author must point out that our message is the most important one maybe in the history of humankind. It is certainly the first period that such dire messages have had to go out. Within the coming century, every nation or governing group all over the world must decide to bring their population forever within the bounds of easy sustainability. They must develop the physical, social and other ways and means to live well but which also ensures that the earth and all its systems can also live well—forever.

An awful prescription! But it must be done. The only alternative is an early extinction.

Acknowledgments

The author acknowledges the great assistance for this book by Brian Clark (grandson). Himself a writer and world traveler, he poured endless hours into revising and improving this book with cogent and necessary deletions, changes, and suggestions on practically every page. These resulted not only in vastly improving the narrative but also the ultimate shrinkage of 3 or 4 dozen pages to boot.

The author also thanks George Sibley, writer and conservationist from Gunnison, Colorado, for his careful, intelligent and very useful changes, alterations and improvements throughout the book. (This, too, resulted, ultimately, in the useful shrinkage of almost 50 pages.)

Special thanks are also due Mike Davidson and the editorial crew at AMZ Book Publishing for their excellent editing and work in pulling this work along

About the Author

Kenneth Deboer was born and raised on a dirt-poor farm in north-central Minnesota. He graduated from the University of Minnesota in 1960 with a BA in Zoology. He went into the Navy in 1961 as a naval aviator, serving in the Pacific, including a short stint in Vietnam. After discharge in 1967, he entered a Ph.D. program in Reproductive Physiology at Iowa State University, graduating in 1970. He then taught various biology courses, including general biology, ecology, genetics, and physiology, at small colleges for almost 20 years. Beginning in 1990, he founded a couple of biotechnology companies at the Advanced Technology Park on the Montana State University campus, where he was an Adjunct Professor. The companies were involved with genetic engineering, with the ultimate goal of using transgenic animal or plant hosts to produce useful drugs and other biological compounds (i.e., Gene Pharming). The main company was based on his and his colleagues' central patent to produce human monoclonal antibodies in transgenic animals, including chickens. This company was sold in 2001, and he semi-retired to a small ranch in central Montana. In 2013, he fully retired to town with his wife and nearby children. Along with a couple of patents and a couple dozen scientific articles, he is the author of four other little books.

REFERENCES

- Ackerman D., 2014. *The Human Age: The World Shaped by Us.* W.W. Norton, NY.

- Ahmed S., Kanger R, 2005. The use of biomass as an alternative energy. Intl. J. Sci. Eng.& Appl. Sci.3(2):58-62.

- Alexander P. et al 2016. Human Appropriation of Land for Food: The Role of Diet. Global Environmental. Change 41:88-98.

- Allen D.L. 1954. *The Legacy of Wildlife.* Funk & Wagnalls, NY.

- Alexander R., Aragon O. et al. 2020. The neuroscience of positive emotions and effects. The implications for cultivating happiness and well-being. Neuroscience Reviews 121:220-249.

- Alpervitz G. 2011. *America Beyond Capitalism: Reclaiming Our Wealth, Our Liberty and Our Democracy.* John Wiley & Sons, NY.

- Amen D. 2020. *The End of Mental Illness.* Tyndale House Publishers, Carol Stream, Illinois.

- Andersen K. 2020. *Evil Geniuses, The Unmaking of America: A Recent History.* Random House, NY.

- Applebaum A. 2020. *Twilight of Democracy: The Seductive Lure of Authoritarianism.* Doubleday, NY.

- Babiak P., Hare R. 2006. *Snakes in Suites: When Psychopaths Go to Work.* Harper Business, NY.

- Badgely C., Perfecto I. 2007. Can organic agriculture feed the world? Renewable Agriculture & Food Systems;22(2):80-85.

- Baille J.L.M., 2010. *Evolution Lost: Status and Trends of the World's Vertebrates.* Zoological Soc. London.

- Bakke G. 2016. *The Grid: The Fraying Wires Between Americans and Our Energy Future.* Bloomsbury, NY.

- Banerjee A, & Duflo E. 2019. *Good Economics for Hard Times.* Hachette Book Group, NY.

- Barbieri M. 2005. *Biosemiotics: A new understanding of Life.* Naturwissenshaften 95(2008), 577-599.

- Barnes P., 2006. *Capitalism 3.0: A guide To Reclaiming the Commons.* Berret-Koehler Publishers, Inc. San Fran.

- Barnes P. 2014. *With Liberty and Dividends for All: How to Save Our Middle Class When Jobs Don't Pay Enough.* Berret-Kohler Pub. Co., San Fran.

- Bartlett A. A. 1998. Reflections on sustainability, population growth, and the environment revisited. Renewable Resources J. 15(4):6-23.

- Bartlett A. A. 2012. Reflections on sustainability and population growth.pp.29-40. IN: Cafaro P. & Crist E. (eds.) *Life on the Brink: Environmentalists Confront Overpopulation.* Univ. Georgia Press, Athens.

- Bar-On Y.M., Phillips R., Milo R. 2018. The biomass distribution on earth. Proc. Nat. Acad. Sci. (2018), (25):6506-6511.

- Beavan Colin, 2009. *No Impact Man: The Adventures of a Guilty Liberal Who Attempts to Save The Planet and The Discoveries He Makes About Himself in The Process.* Farrar, Strauss & Giroux, NY.

- Beckoff M. 2001. Social play, behavior, cooperation, fairness, trust, and the evolution of morality. Conservation Studies 8:81-91.

- Belanger J.D., Pillings D. (Eds). 2019. The State of the World's Biological Biodiversity for Food and Agriculture. FAO Commission on Genetic Resources for Food and Agriculture.

- Berry T. 1999. *The Great Work: Our Way into The Future.* Bell Tower, NY.

- Berry W. 2003. *Citizenship Papers.* Shoemaker & Hoard, NY.

- Berry W. 1996. *The Unsettling of America; Culture & Agriculture.* Sierra Club Books, San Francisco.

- Bish J. 2012. Toward A New Armada: A Globalist Argument for Stabilizing the U.S. Population, p.189-201. IN: Cafaro P. & Crist E. (Eds). *Life on The Brink: Environmentalists Confront Overpopulation.* Univ. Georgia Press, Athens, GA.

- Blewett J., Cunningham R., (eds.). 2014. *The Post-Growth Project: How the End of Economic Growth Could Bring a Fairer and Happier Society.* Green House, London.

- Bradshaw C. B. Brooks, 2014. Human population reduction is not a quick fix for environmental problems. Proc. Nat. Acad. Sci. 111(46):166610-15.

- Brandin J.M., Turner I. & Odenbrand. 2011. *Small-scale Gasification: Gas Engine to Liquid Biofuel.* Linnaeus Univ. Varjo, Sweden.
- Bregman, R. 2017. *Utopia for Realists: How We Can Build the Ideal World.* Little, Brown, & Co., Boston.
- Brennan R., Garret A.D., Huber K., Hargarten H., Pespeni M. Rare genetic variation and balanced polymorphisms are important for survival in global change conditions. https://doi.org/10.1098/rspb.2019.0943
- Bricker D. and Ibbitson J. 2019. *Empty Planet: The Shock of Global Population Decline.* Crown Publishing Group, NY.
- Bromfield L. 1945. *Pleasant Valley.* The Wooster Book Co., Wooster, Ohio.
- Brown L. 1974. *In the Human Interest: A Strategy for Stabilizing World Population.* W.W. Norton, NY.
- Brown L. 2008. *Plan B 3.0: Mobilizing to Save Civilization.* W.W. Norton, NY.
- Brown L. 2011. *World on The Edge: How to Prevent Environmental and Ecological Collapse. Earth Policy Institute.* W.W. Norton, NY.
- Brown L. 2012. Environmental refugees. Pp 108-122. IN: Cafaro P. & Crist E. (eds): *Life on The Brink: Environmentalists Confront Overpopulation.* Univ. Georgia Press, Athens, GA.
- Bruffe K.A. 1999. *Collaborative Learning: Higher Ed, Interdependence, and the Authority of Knowledge.* 2nd Ed, Johns Hopkins Univ Press, Baltimore, MD.
- Burkhead N. M. 2012. Extinction rates in North American freshwater fishes, 1900-2010. Bioscience 62(9):798-808.

- Campbell M. 2012. Why the silence on population? Pp 41-55. In: Cafaro P. & Crist E. (eds). *Life on the Brink: Environmentalists Confront Overpopulation.* Univ. Georgia Press, Athens, GA.

- Cafaro P. & Crist E. (Eds.) 2012. *Life on The Brink: Environmentalists Confront Overpopulation.* University of Georgia Press, Athens.

- Cafaro P. & Staples, 2012. The environmental argument for reducing immigrations into the United States, pp 182-188. IN: Cafaro P. & Crist E. (eds). *Life on the Brink: Environmentalists Confront Overpopulation.* Univ. Georgia Press, Athens, GA.

- Carpenter K.E. et al. 2008. One-third of reef-building corals face elevated extinction risk from climate change and local impacts. Science 321(5888:56-563).

- Carson R. 1962. *Silent Spring.* Houghton Mifflin, Boston

- Carter V.G. & Dale T. 1974. *Topsoil and Civilization.* Univ. Okla Press, Norman.

- Catton, Wm. 2012 Destructive Momentum: Could an Enlightened Environmental Movement Overcome It? pp 16-28. IN: Cafaro P. & Crist E. (Eds) *Life on the Brink: Environmentalists Confront Overpopulation.* Univ Georgia Press, Athens, GA.

- Catton, W. 1982. *Overshoot: The Ecological Basis of Revolutionary Change.* Univ Ill. Press, Urbana, Ill.

- Caviccioli R., Ripple W.J.R. & Webster N.S. 2019. Scientists warning to humanity: Microrganisms and climate change. Nature Reviews, Microbiology 17:569-586.

- Chen Y. et al. 2013. Evidence on the impact of sustained exposure to air pollution on life expectancy from China's

Huai River Policy. Proc. Nat. Acad. Sci. 110:12936-41.

- Chiavani E. & Bernstein A. (Eds). 2008. *Sustaining Life: How Human Health Depends on Biodiversity.* Oxford Univ Press, NY.

- Christian E.J. 2012. *Seed Development and Germination of Miscanthus sinensis.* Dissertation, Iowa State Univ.

- Christensen C. 2011. *The Innovator's Dilemma.* Harper Business, NY.

- Churchland P. 1995. *The Engine of Reason, the Seat of the Soul: A Philosophical Journey into the Brain.* MIT Press, Cambridge, Mass.

- Clover C. 2006. *The End of the Line: How Overfishing Is Changing the World and What we Eat.* The New Press, NY.

- Cohen J. 2015. *How Many people Can the World Support?* W.W. Norton, NY.

- Collins P. *The Depopulation Imperative: How Many People Can the Earth Support?* Australia Scholarly Publ, Sidney.

- Crary J. 2020. *Scorched Earth: Beyond the Digital Age to A Post-Capital World.* Verso, NY.

- Crist E. Limits to Growth and the Biodiversity Crisis. Wild Earth, Spring 2003,63.

- Crist, E. 2012. Abundant Earth and the Population Question., pp. 141-153, IN: Cafaro P. & Crist E. (eds). *Life on The Brink: Environmentalists Confront Overpopulation.* Univ Georgia Press, Athens, GA.

- Crutzen P. Geology of Mankind. Nature, 415, 23 (2002): https://doi.org/10.1038/4150230.

- Czech B. 2000. *Shoveling Fuel for A Runaway Train.* U. Cal. Press, Berkeley.

- Dailey G., Ehrlich A. & Ehrlich P. 1994. Optimum Human Population Size. Population. & Environ: A Journal of Interdisciplinary Studies 16(6):469-475.

- Daly H. 1977. *Steady-state Economics: The Economics of Biophysical Equilibrium and Moral Growth.* W.H. Freeman, San Fran.

- Daly H, 1996 *Beyond Growth: The Economics of Sustainable Development.* Beacon Press, Boston.

- Daly, H. 2007. *Ecological Economics and Sustainable Development: Selected Essays of Herman Daly.* Edward Elgar, Cheltenhan, U.K.

- Daly, H. & J. Farley. 2003. Steady-state Economics: Principles and Applications. Island Press, Wash, DC.

- Damasio A. 2010. *Self Comes to Mind: Constructing the Conscious Brain.* Vintage Books, NY.

- Dee T. 2015 *Four Fields.* Counterpoint Press, Berkeley,

- Dennett D. 2018. *From Bacteria to Bach and Back: The Evolution of Minds.* W.W. Norton, NY.

- Dewaal F. 2019. *Mama's Last Hug: Animal Emotions and What They Tell Us About Ourselves.* W.W. Norton, NY.

- Diamond J. 2012. *The World Until Yesterday: What We Can Learn from Traditional Societies?* The Penguin Group, NY.

- Diesing P. 1962. *Reason in Society.* Univ. Illinois Press, Urbana.

- Dirzo R. et al. 2014. Defaunation in the Anthropocene. Science; 345:401-406.

- Dittmar H, et al. 2014. The relationship between materialism and personal well-being.: A meta-analysis. J. Personal and Social Psychol. 107:8879-924.

- Dodson L. 2008. *The Moral Underground: How Ordinary Americans Subvert an Unfair Economy.* New York Press,

- Doidge, N. 2015. *The Brain's Way of Healing: Remarkable Discoveries and Recoveries from the Frontiers of Neuroplasticity.* Viking, NY.

- Egan, T. 2006. *The Worst Hard Time: The Untold Story of Those Who Survived the Great American Dust Bowl.* Houghton Miffllin, Boston.

- Ehrenfeld D. 1978. *The Arrogance of Humanism.* Oxford Univ. Press, NY.

- Ehrlich P., Ehrlich, A. 1968. *The Population Bomb.* Ballantine Books, NY.

- Ehrlich P., Ehrlich A. 1981, *Extinction: The Causes and Consequences of the Disappearance of Species.* Random House NY.

- Eldridge N. 1998. *Life in The Balance: Humanity and the Biodiversity Crisis.* Princeton Univ. Press, Princeton, NJ.

- Elliott, H. 2005. Ethics For a Finite World: An Essay Concerning A Sustainable Future. Fulcrum, Golden, Colo.

- Emmott S. 2013. *Ten Billion.* Random House, NY.

- Englemann R. 2008. *More: Population, Nature, and What Women Want.* Island Press, Wash, DC.

- Englemann R. (2012. Trusting Women to End Population Growth, pp 223-239, IN: Cafaro P. & Crist E. (Eds). *Life on The Brink: Environmentalists Confront Overpopulation.* Univ. Georgia Press, Athens.

- Erle E., Ramankthu N. 2018. Geologic Society of America Report on U.S. Conditions.

- Esterlin R., A. McVey et al. 2010. The happiness-income paradox revisited. Proc. Nat. Acad Sci. 107(52):22463-22468.

- Estes J.A. et al. 2011. Trophic downgrading of planet Earth. Science333(6040):301-306.

- European Photovoltaic Ind. Assoc. 2010. Roofs could technically generate 40% of EU's electricity demand by 2020, (Press Release, June). From; Http://www.org/fileadmin/EPIA_docs/;public/100623_PR_B IPV_En.pdf

- Ewing B.S. et al. 2009. The Ecological Footprint Atlas. www.footprintnetworkk.org

- Falk B. 2013. *The Resilient Farm and Homestead: An Innovative Permaculture & Whole System Design Approach.* Chelsea Green Publishing, White River Junction, Vermont.

- Famiglietti J.S. 2014. The global groundwater crisis. Nature: Climate Change 4:945-948.

- Feinstein J.S. et al. 2011. The human amygdala and the induction and experience of fear. Current Biol. 21:34-38.

- Fishman C. 2011. *The Big Thirst: The Secret Life of Water and Turbulent Future of Water.* Free Press (Simon & Schuster) NY.

- Foreman D. (with Laura Carrol). 2014. *Man Swarm: How Overpopulation is Killing the Wild World. Second Edition*, LiveTrue Books, NY.

- Frey G.B. 2015. The end of economic growth. Sci. Am 312(1):12.

- Friedman G. 2009. *The Next 100 Years: A Forecast for the Next Century.* Doubleday, NY.

- Freidman T. 2005. *The World Is Flat: A Brief History of the Twenty-First Century.* Farrar, Strauss & Giroux, NY.

- Friend T. 2019. Value Meal. The New Yorker, Sept 30,42-55.

- Galbreath J.K. 1958. *The Affluent Society*, Houghton Mifflin, Boston.

- Gates B. 2021. *How to Avoid A Climate Disaster: The Solutions We Have and the Breakthroughs We Need.* Alfred A. Knopf, NY.

- Gates Robert. 2020. *Exercise of Power: American Failures, Successes, And A New Path Forward in The Post-Cold War World.* Penguin Random House, NY.

- Gazzaninga M. 2018. *The Consciousness Instinct: Unraveling the Mystery of How the Brain Makes the Mind.* Farrar, Straus & Girous, NY.

- Gilbert D. 2006. *Stumbling on Happiness.* Alfred A. Knopf, NY.

- Giridharadas A. 2018. *Winners Take All: The Elite Charade of Changing the World.* Vintage Books, NY.

- Glavin T, 2007. *The Sixth Extinction: Journeys Among the Lost and Left Behind.* St Martin's Press, NY.

- Gleick P.H., & Palaniappan. M. 2010. Peak water limits to freshwater withdrawal and use. Proc..Nat. Acad. Sci. 107:11155-62.

- Godfrey H.C.J. Et al 2010. Food security: The challenge of feeding 9 billion people. Science 327:812-918.

- Goleman D. 2009. *Ecological Intelligence: How Knowing the Hidden Costs of What We Buy Can Change Everything.* Doubleday (Random House), NY.

- Goodell J. 2010. *The Water Will Rise: Rising Seas, Sinking Cities and Remaking of the Civilized World.* Little, Brown & Co., NY.

- Grandin G. 2019. *The End of The Myth: From the Frontier to The Border Wall in The Mind of America.* Henry Holt & Co. NY.

- Grant, L. 2006. *The Case for Fewer People.* The NPG Forum Papers. Seven Locks Press, Santa Ana, Ca.

- Greene J., 2013. *The Moral Tribe: Emotion, Reason and the Gap Between US and THEM.* The Penguin Press, NY.

- Gross B. 2010. "Private Eye". Investment Overlook (Blog), PIMCO, Your Global Investment Authority. www.pimco.com/En/Insights/pages?PrivateEyeBillGrossAugust2010.aspx.

- Gu, H., Bergman R. 2017. Cradle-to-grave life cycle assessment of syngas electricity from woody biomass refuse. Wood & Forest Science, 49(2):177-192.

- Haberl, H., Erb K., Krausmann F. 2010. Global Human Appropriation of Net Primary Production (HANPP). Encyclopedia of the Earth, www.eoearth.org

- Haglund, M. 2019. *This Life: Secular Faith and Spiritual Freedom.* Panteon Books, Knopf Doubleday Pub. NY.

- Halladay, T. 1978. *Vanishing Birds: Their Natural History and Conservation.* Rinehart & Winston, NY.

- Haiming, J., Larson E., Celik F. 2009. Modelling and analysis: Gasification-based electric power generation from switchgrass. Biofuels, Bioprod., Bioref. 3:142-173.

- Hardin, G. 2014, An ecolate view of the Human Predicament. HTTP://garretthardinsociety.org./articles/art_ecolate_view_human_predicament.html

- Hare, B., Kweduenda S.K. 2010. Bonobos voluntarily share their own food with others. Current Biol. 20:230-231.

- Hardin, G. 1993. *Living Within Limits: Ecology, Economics and Population.* Oxford U Press, NY.

- Hawken, P. 2017. *Drawdown: The Most Comprehensive Plan Ever Proposed to Reverse Global Warming.* Penguin Books, NY.

- Hawkins, R. 2012, Perceiving overpopulation: Can't we see what we are doing? Pp 202-213. In: Cafaro P. & Crist E. (eds). *Life on the Brink: Environmentalist Confront Overpopulation.* Univ. Georgia Press, Athens, Ga.

- Hawkins, T. et al 2013. Comparative environmental life cycle assessment of conventional and electric vehicles. J. of Industrial Ecol. 17:53-66.

- Hedges, C. 2009. *Empire of Illusion: The End of Literacy and the Triumph of Spectacles.* Nation Books, NY.

- Hedges, Chris. 2015. *America: The Farewell Tour.* Simon & Schuster, NY.

- Heinberg. R. 2010. *Peak Everything: Waking Up to the Century of Declines.* New Society Publ, Gabriola Island, Brit. Columbia, Canada.

- Heinberg, R. 2011. *The End of Growth: Adapting to Our New Environmental Reality.* New Society Publishers, Gabriola Island, B.C.

- Heinberg, R. 2021. *Power: Limits and Prospects for Human Survival.* New Society Publ, Gabriola, BC, Canada.

- Henderson H. 2006. *Ethical Markets: Growing the Green Economy.* Chelsea Publ.

- Herrero, M., Thornton P.K. 2013. Livestock and global change: Emerging issues for sustainable food systems. Proc. Nat. Acad. Sci.110(52):20878-20881.

- Hibberd, J. et al. 2008. Using C4 photoynthesis to increase the yield of rice---Rational and feasibility. Curr. Opinions Plant. Biol. 11:228-231.

- Hibbing J.R., Alford J. & Smith K.B. 2013. *Predisposed: Liberty and Conservatives and the Biology of Political Differences.* Routledge Publishing, NY.

- Hicks J., Labell D., Asner G. 2004. Cropland area and net primary productivity computed from 30 years USDA harvest data. https://doi.org/10.1175/1087-3562(2004)008<001:CAANP>2.0c0.2

- Hoekstra, A. & Wiedman T. 2014. Humanity's unsustainable footprint. Sci. 344(6188):1114-1117.

- Hughes, T. et al. 2003. Climate change, human impacts, and the resilience of coral reefs. Science 301:929.

- Integrated Information Networks. 2002. Pakistan: Focus on Water Crisis. IRIN News, May 17.

- Imhoff, M.L. et al 2004. Global patterns in human consumption of net primary production. Nature 2004:429(6994):870-873.

- Irvine, S. The Great Denial: Puncturing Pro-natalist Myths. Wild Earth, Winter, 1997/1998,8.

- Jackson, J.B.C. 2008. Ecological extinction in the brave new ocean. Proc. Nat. Acad. Sci. 105(suppl.1):11458-11465.

- Jackson, J.B.C., Kirby M.X. et al. 2001. Historical overfishing and the recent collapse of coastal ecosystems. Science 293(21),629-638.

- Jackson, T. 2017. *Prosperity Without Growth: Foundations for The Economy of Tomorrow.* Routledge, NY.

- Jasanoff A. 2018. *The Biological Mind: How Brain, Body and Environment Collaborate to Make Us Who We Are.* Basic Books, NY.

- Jensen, D. 2006. *Endgame: Vol. 1. The Problem Of Civilization & Vol.2. Resistance.* Seven Stories Press, NY.

- John, S. & Watson A. 2007. Establishing Grass Energy As A Crop Market in the Decatur Area. The Agricultural Watershed Institute, Univ. Illinois.

- Johnson, K. Northern Long-Eared Bat Down 99% due to white nose. Defenders of Wildlife 95(1):18, 2019.

- Jonkers L, Hillebrand H., Kucers M. 2019. "Global changes drives modern plankton communities away from the pre-industrial state. Nature (May, 2019).

- Joshua T. and Vogelstein et al. 2014. Discovery of brain wide neural-behavioral maps via multiscale unsupervised structure learning. Science (344(2014):386-392.

- Jost J. et al. 2014. Political neuroscience: The beginning of a beautiful friendship. Adv. In Pol. Psychology 35; Suppl.1.

- Justin S., Feinstein et al. 2011. The human amygdala and the induction and experience of fear. Current Biology 21 (2011),34-38.

- Kahneman, D. & Kreuger A. 2006. Developments in the measurement of subjective well-being. J. Econ. Perspectives 20(1):3-24.

- Kahneman, D & Deaton A. 2010. High income improves evaluation of life but not emotional well-being. Proc Nat. Acad. Sci. 107:16489-92.

- Kanase-Patil, A.B., Saini R.P. and Sharma M.P. 2007. Biomass based electrical production technologies. Natl. Convention of Chem. Engineers in "Recent Trends in Chemical Engineering, Oct. 2007.

- Kanai R., Felden T., Firth C. & Ries G. 2011. Political orientations are correlated with brain structures in young adults. Curr. Biol 21:677-680.

- Kassar T. 2008. *A Vision of Prosperity.* London Sustainable Development Commission.

- Kassar T. 2002. The High Price of Prosperity. MIT Press, Cambridge, MA.

- Kelly, M. 2009. *The Divine Right of Capital.* Barrett-Koehler, SF.

- Kimmerer, R. W. 2013. *Braiding Sweetgrass: Indigenous Wisdom, Scientific Knowledge, and the Teachings of Plants.* Milkweed Editions, Canada.

- Klein, N. 2014. *This Changes Everything.* Simon & Schuster, NY.

- Klein, N., 2017. *No Is Not Enough: Resisting Trump's Shock Politics and Winning the World We Need.* Haymarket Books, Chicago.

- Kolbert E. 2009. *The Sixth Extinction: An Unnatural History.* Henry Holt, NY.

- Kolandkiewicz L and Beck R. 2001. Weighing sprawl factors in large U.S. cities. Numbers USA. https://www.lnubersusa.com/content/files/LargeCity%Sprawl.pdf.

- Kolandkiewicz L. 2012. Overpopulation versus biodiversity, pp 75-90. IN: Cafaro P. & Crist E., (eds). *Life on the Brink: Environmentalists Confront Overpopulation.* Univ. Georgia Press, Athens.

- Konczal, M. 2021. *Freedom from the Market: America's Fight to Liberate Itself from the Grip of the Invisible Hand.* The New Press. NY.

- Korten D. 1999. *The Post-Capital World: Life After Capitalism.* Berret-Koehler Pub., NY.

- Kotsko, A. 2012. *Neoliberal Demons: On the Theology of Late Capital.* Stanford U. Press, Palo Alto.

- Kozlowski, T. 1980. Impacts of air pollution on forest ecosystems, Bioscience 30:88-93.

- Krugman, P. 2020. Americans Living the Republican Dream, New York Times, 27 July.

- Kunstler, J. 2019. Foreword, to Wallace-Wells "The Uninhabitable Earth."

- Lakoff, G. 2002. Moral Politics: How Liberals and Conservatives Think. Univ. Chicago Press, Chicago.

- Lanchester J. 2019. The Invention of Money. The New Yorker, August 5 & 12, p.28-31.

- Lang. S. 2006. *Slow, insidious soil erosion threatens human health and welfare as well as the environment.* Cornell Study, March, 2006.

- Lawton, K. Economics of soil loss; In: Farm Progress Report, March 13, 2-17.

- Leakey R. & Lewin R. 1977. *Origins.* E.P. Dutton, NY

- Leakey R., and Lewin R. 1996. *The Sixth Extinction: Patterns of Life and the Future of Mankind.* Anchor Books, NY.

- Lelieveld, L. et al. 2015. The contribution of outdoor air pollution sources on premature mortality on a global scale. Nature 525:367-371.

- Lovins, A. 1979. *Introducing the Soft Environmental Paths Towards A Durable Peace.* Harper & Row, NY.

- Leopold A, 1949. *The Sand County Almanac: And Sketches Here and There.* Oxford U. Press, NY.

- Lowdermilk, W.C. 1935. Conquest of the Land Through 7,000Years. Ag. Info. Bull. No. 99, US Dept Agriculture, August, 1953.

- Lowery A. 2018. *Give People Money: How A Universal Basic Income Would End Poverty, Revolutionize Work and Remake the World.* Crown, NY.

- Lu X., et al. 2019. Gasification of coal and biomass as a net carbon-negative power source for environmentally-friendly electricity. Pros. Nat. Acad. Sci. 116(10):8206-8213.

- MacLean, N. 2017. *Democracy in Chains: The Deep History of the Radical Right's Plan for America.* Penguin Random House, NY.

- Mann, C. 2018. *The Wizard & The Prophet: Two Remarkable Scientists and Their Dueling Visions to Shape Tomorrow's World.* Alfred A. Knopf, NY.

- Marcus, R. 2019. *Supreme Ambition: Brett Kavanaugh and the Conservative Takeover.* Simon & Schuster, NY.

- Manning, R. 2004. *Against the Grain: How Agriculture Has Hijacked Civilization.* Farrar, Strauss & Giroux, NY.

- Marsh, C. P. 1867, *Man and Nature.* C. Scribner, NY.

- Mason, P. 2015. *Post capitalism: A Guide to the Future.* Farrar, Straus & Giroux. NY.

- Matsutani, A. 2006. Shrinking-population Economics: Lessons from Japan. Intl. House of Japan, Tokyo.

- McCauley, K.J. et al 2015. Marine defaunation: Animal loss in the global ocean. Science347(6219):247-254.

- McCoy, A. Alter Net, Dec. 16, 2021

- McNeill W. 1976. *Plagues and Peoples.* Anchor Press/Doubleday, NY.

- Mazur, L. 2009. *A Pivotal Moment; Population, Justice and the Environmental Challenge.* Island Press, Wash, DC.

- McKee J.K. 2005. *Sparing Nature: The Conflict Between Human Population, Growth and Earth's Biodiversity.* Rutgers U Press, New Brunswick, NJ.

- McKibben B., 1989. *The End of Nature.* Anchor Books, NY.

- McKibben, Bill. 2007. Deep Economy: The Wealth of Communities and the Durable Future. Macmillan, St. Martin Publ. Group, NY.

- McKibben, B. 2019. *Falter: Has the Human Game Begun to Play Itself Out?* Henry Holt & Co., NY.

- Meadows D., Meadows D., Randers J., Behrens W.W. 1972. *The Limits to Growth.* Club of Rome, Potomac Associates, NY.

- Millenium Ecosystem Assessment. Synthesis 2005. www.milleniumassessment.org

- Mekonnen M. & Hoekstra A.Y. 2016. Four Billion People Facing Severe Water Scarcity. Science Advances 2(2): e1500233.

- Mill, J. S. 1848. *Principles of Political Economics and Liberty.* (from F. Ashley Library, 1909).

- Mishra, P. 2017. *The Age of Anger: A History of the Present.* Farrar, Strauss & Giroux. NY.

- Mitchell, A. 2009. *Seasick: Ocean Change and Extinction of Life on Earth.* Univ. Chicago Press, Chicago, 2009.

- Montgomery, D. 2001. *Dirt: The Erosion of Civilization.* Univ. Calif Press, Berkeley.

- Montgomery, D. 2007. Soil erosion and agricultural sustainability. Proc. Nat. Acad. Sci. 104(33):13268-72.

- Mooney, C. 2012. *The Republican Brain: The Science of Why They Deny Science—and Reality.* Wiley & Sons, NY.

- Moynihan T. 2020. *X-Risk: How Humanity Discovers Its*

Own Extinction. Urbanomic Medig LTD, Falmouth, UK.

- Mumford L. 1956. *The Transformation of Man.* Harper & Row, NY.

- Murtaugh, P, Schlax M. 2009. Reproduction and the Carbon Legacy of Individuals. Global Environmental Changes 19(1):14-20.

- Mussel, P, Hewig J. 2019. A neural perspective on when and why trait greed comes at the expense of others. Sci. Reports 9:10985-91.

- Myers, R. Worm B. 2003. Rapid worldwide depletion of predatory fish communities. Nature 423(2003):280-283.

- Nash, R. 2012. Island Civilization: A Vision for Human Inhabitance in the Fourth Millenium. PP 301-327, IN: Cafaro P., Crist E. (eds). Life on the Brink: Environmentalists Confront Overpopulation. Univ. Georgia Press, Athens.

- NasaJPL. 2015. Water Storage. www.gracefo.jpl.nasa.gov.science.water-storage/

- Narvaez, R.A., et al 2018. Low-cost syngas shifting for remote gasifiers: Combustion of CO_2 adsorption and catalyst additives for novel and simplified packed structures. Energies 2018, 11:311.

- Nelson, V. & Starcher K. 2017. *Introduction to Bioenergy.*, CRC Press. NY.

- Newitz, A. 2013. *Scatter, Adapt and Remember: How Humans Will Survive A Mass Extinction.* Doubleday, NY

- Opie, J. 1993. *Ogallala: Water on A Dry Land.* Univ. Nebraska Press, Lincoln.

- Opie, J., C. Miller and K. Archer, 2018. 3rd Ed Ogallala:

Water On A Dry Land. Univ. Nebraska Press, Lincoln.

- Ortman, J.M., & Guarneri C.E. 2009. United States Population Projections: 2000 to 2050. U.S. Census Bureau.

- Osborn, F. 1948. *Our Plundered Planet.* Little, Brown, Co. Boston.

- Pahl, G. 2007. *The Citizen-powered Energy Handbook: Community Solutions to a Global Crisis.* Chelsea Green Publishing Co, White River Junction, VT.

- Palmer, P. J. 2011. *Healing the Heart of Democracy: The Courage to Create A Politics Worthy of the Human Spirit.* Jossey-Bass, San Francisco.

- Palmer T. 2013. Beyond futility. Pp 98-107, In: Carfaro P. & Crist E., (Eds). *Life on The Brink: Environmentalists Confront Overpopulation.* Univ. Georgia Press, Athens.

- Panskeep, J. & L. Biven. 2012. *The Archeology of Mind: Neuro-evolutionary Origins of Human Emotion.* WW. Norton, NY.

- Pauley, D et al. 2002. Toward sustainability in world fisheries. Nature 418(2002):689-695.

- Pantaleo, AM. Et al. 2012. Bioenergy routes for heat and power in urban areas: Trade-offs and perspectives for smart cities. Biomass & Bioenergy doi.10.1016/j.biomioe.2012.0.022

- Pereora H.M. et al. 2013. Essential biodiversity variables. Science 339 (6117):278-279.

- Peterson, S.M. et al. 2016. Groundwater-flow Model of The Northern High Plains Aquifer in Colorado Kansas, Nebraska, South Dakota & Wyoming. US Geological Survey Scientific Investigations Report 2016-5153. USGS

- Philpott, T. 2020. *Perilous Bounty: The Looming Collapse of American Farming and How to Prevent It.* Bloomsbury Publishing, NY.

- Pianki, Eric. 2006. The Vanishing Book of Life On Earth. Zo.utexas.edu/courses/bio3731/Vanishing.Book. pdf.

- Piketty, T. 2019. *Capital and Ideology.* Harvard Univ. Press, Cambridge.

- Pimentel, D. 2006. Soil erosion: A Food and environmental threat. Environ. Devel. Sustain.: 8,119.

- Pimm, S.L., Ayres M., et al. 2001. Can we defy natures' end? Science 293(2001), 2207-2208.

- Pimm, S.L. et al. 2014. The biodiversity of species and their rates of extinction, distribution and protection. Science 344(6187):12467520-10.

- Pinker, S. 2011. *The Better Angels of Our Nature: Why Violence Has Declined.* Penguin Viking Books, NY.

- Pinker, 2018. *Enlightenment Now: The Case for Reason, Science, Humanism, and Progress.* Viking, Penguin Random House, NY.

- Potts, M. 1997. Sex and the Birth Rate: Human Biology, Demographic change, and Access to Fertility-regulation Methods. Population & Development Reviews 23(1):1-39.

- Ponting, C. 1991. *A Green History of the World.* St. Martins, Pres, NY.

- Prata, N. 2009. *Making Family Planning Accessible in Resource-poor Settings.* Philos. Trans Roy. Soc, B364:3093-3099.

- Punke, M. 2007. *Last Stand: George Bird Grinnell, The Battle to Save The Buffalo, and the Birth of the New West.* Harper Collins Pub, Inc. NY.

- Putnam, R. 2001. *Bowling Alone: The Collapse and Revival of American Community.* Simon & Schuster, NY.

- Putnam, R. *The Upswing: How America Came Together A Century Ago and How We Can Do It Again.* Simon & Schuster, NY.

- Rakesh, K. *10 Projections for the Global Population in 2050.* PEW Research FactTank, Feb, 2014.

- Raworth, K. 2017. *Donut Economics: Seven Ways to Think Like a 21st Century Economist.* Random House Business, NY.

- Rees, W.E., & M. Wackernagel, IN: Jansson A.M., et al (eds). Investing in Natural Capital; The Ecological Economics, 1994. Island Press, Wash. DC.

- Reich, R. 2015. *Saving Capitalism for the Many, Not the Few.* Alfred A. Knopf, NY.

- Reich, R. 2018 *The Common Good.* Alfred A. Knopf NY.

- Reich, R., 2020. *The System: Who Rigged It, How We Fix It.* Knopf Doubleday Pub. Group, NY.

- Reid, W. et al. 2005. *Ecosystems and Human Well-Being: Synthesis.* Millennium Ecosystem Assessment series, Island Press, Wash. DC.

- Reisner, M. 1986, 1993, Postscript 2017 with Lawrie Mott. *Cadillac Desert: The American West and Its Disappearing Water.* Penguin Random House, NY.

- Rich, N. 2019. *Losing Earth: A Recent History.* McMillan, NY.

- Rifkin J. 2011. *The Third Revolution: How Lateral Power is Transforming Energy, the Economy, and the World.* Palgrave Macmillan, NY.
- Roberts, Paul. 2004. *The End of Oil: On the Edge of a Perilous New World.* Houghton Mifflin Co., NY.
- Roberts, P. 2008. *The End of Food.* Houghton Mifflin Harcourt Pub. Co., NY.
- Rosenberg, K. et al. 2019. Decline of the North American avifauna. Science 366(6461):120-124.
- Ryerson, W. 2012. How do we solve the population problem? Pp 240-254. IN: Cafaro P. & Crist E. (eds). *Life on The Brink: Environmentalists Confront Overpopulation.* Univ. Georgia Press, Athens.
- Sachs, J. 2005. *The End of Poverty: Economic Possibilities for Our Times.* Penguin Press, NY.
- Sala, O., Chapin F.S., et al. 2000. Global diversity scenarios for the year 2100. Science 287 (2000), 1770-1774.
- Schlesinger, Arthur, Jr. 1957(Rev. 2002). The Crisis Of The Old Order. Houghton Mifflin, Boston.
- Schlesinger, A., Jr. 1960. *The Age of Roosevelt: The Politics of Upheaval.* Houghton Mifflin, Boston.
- Schneider, K. 2019. *The U.S. needs a Mental Health Czar.* Sci. Am., Nov, 2019.
- Sears, P. 1980. *Deserts on the March.* Univ. Okla. Press, Norman, OK.
- Searles, J. 2010. *Making the Social World.* Oxford U. Press, Oxford, UK.

- Sheppard S., Davy K., Pilling D.M. 2009. *The Biology of Coral Reefs.* Oxford Univ Press, Oxford.

- Sherman, J. 2020. This Is the Rhetorical Trick Most Likely to Cause the Extinction of Our Species. Alternet.org., 20 July, 2020.

- Shragg, K. 2015. *Move Upstream: A Call to Solve Overpopulation.* Freethought House, Minneapolis-St Paul.

- Sibley, G. 2022. Romancing the River; Pts. 1-6. HTTPS://sibleysrivers.com

- Siegel, D. 2017. *The Mind: A Journey to The Heart of Being Human.* WW Norton, NY.

- Sifferlin, A. 2017. Here's How Happy Americans Are Right Now. Harris Annual Poll. Time Magazine, July 2017.

- Simberloff, D. 2013. *Invasive Species: What Everyone Needs to Know.* Oxford U. Press, NY.

- Singh, S., Darroch J.E., Ashford L.S., Vlassoff M. 2009. *Adding It Up; The Cost and Benefits of Investing in Family Planning and Maternal and Newborn Health.* Guttmacher Instit. & UN Population Fund, NY.

- Smaje, C. 2020. *Small Farm Future: Making the Case for a Society Built Around Local Economics, Self-Provisioning, Agricultural Biodiversity and a Shared Earth.* Chelsea Green Publ.

- Smail, J.K. 1997. *Beyond Population Stabilization: The Case for Dramatically Reducing Global Human Numbers. Politics & Life Sciences,* 16, no. 2. Sept, 1997, Beech Tree Publishing, Surrey, UK.

- Smail, J.K. 2002. *Confronting a surfeit of people: Reducing Global Human Numbers to Sustainable Numbers.* Envir.

- Devel. & Sustain. 4, July, 20022, Kluwer Academic Press, Netherlands.

- Sokhansanj, S. et al. 2009 Large-scale production, harvest and logistics of switchgrass (*panicum virgatum*): Current technology and envisioning a mature technology. Biofuels, Bioprod., Biorefining. 3:124-141.

- Speidel, J. et al. 2007. *Family planning and reproductive health: The link to environmental reservation. Bixby Center for Reproductive Health Research and Policy*, Univ. Cal. San Fran.

- Speth, G., 2009. *The Bridge at the Edge of the World: Capitalism, The Environment and Crossing from Crisis to Sustainability.* Yale Univ Press, New Haven.

- Staples W., Cafaro P. 2012. For A Species Right to Exist. Pp 283-300 IN: Cafaro P. & Crist E. (eds), *Life on the Brink: Environmentalists Confront Overpopulation.* Univ Georgia Press, Athens, Ga.

- Stein, B.A., Kutner L.S., Adams J.S. (eds) 2000. *Precious Heritage: The Status of Biodiversity in the United States.* Oxford U Press, Oxford, UK.

- Stuart S.N. et al 2004. Status and trends of amphibian declines and extinctions worldwide. Science 306(5702):1783-1786.

- Sutherland, R. 2016. The dematerialization of consumption. EDGE, htttpe://www.edge.org/response-detail/26750.

- Tekale A. et al 2017. Energy production from biomass. Intl. J. Innovative Sci. & Res. Technology 2(10):26-29.

- Trexler, D. 2012. *Becoming Dinosaurs: A Prehistoric Perspective on Climate Change Today.* Sweetgrass Books, Helena, MT.

- Tsao, Lewis J., Crabtree N. G., Eds. 2006. Solar FAQ's. U.S. Dept Energy, Office of Basic Science (BES).

- Tuschmann, A. 2013. *Our Political Nature: The Evolutionary Origins of What Divides Us.* Prometheus Books, NY.

- Udall, S., Conconi C. & Osterhout D. 1974. *The Energy Balloon.* McGraw Hill, NY.

- UN Dept Economic & Social Affairs (UNDESA). 2011. World Population Prospects: The 2010 Revision. New York; UN.

- Utuk, I.O. & Daniel E. 2015. Land Degradation: A threat to food security: A global assessment. Science (2015) 5:13-21.

- Venter J.C. 2013. *Life at the Speed of Light: From the Double Helix to The Dawn of Digital Life.* Viking Press, NY.

- Verso, E. 2015. Topsoil Erosion. Thesis, Stanford University.

- Victor, P. 2007. *Managing Without Growth: Slower by Design, Not Disaster.* Edward Elgar Pub, Cheltenham, UK.

- Vitousok, PM et al. 1997. Human Domination of earth's ecosystems. Science,1997(5325):494-499.

- Vogt, W. 1948. *The Global Perspective.* IN: Nash R. (Ed). *American Environmentally. Readings in Conservation History.* Third Ed. 1990. McGraw-Hill, NY.

- Vogt, W. 1948. *The Road to Survival.* William Slane Associates, NY.

- Vollman, W. 2018. *Carbon Ideologies: Vol. 1. No Immediate Danger.* Viking: Penguin Random House. NY.

- Vollman, W. 2018. Vol 2. *No Good Alternative.* Viking: Penguin Random House, NY.

- Vollrath, D. 2020. *Fully Grown: Why A Stagnant Economy Is a Sign of Success.* Univ. Chicago Press, Chi.

- Wagg, C. et al 2014. Soil biodiversity and soli community composition determine ecosystem multifunctionality. Proc. Nat. Acad. Sci. 111(14):5266-5270.

- Walsh, D. 2020. *The Nine Lives of Pakistan: Dispatches from a Precarious State.* Bloomsbury Publishing, London.

- Wallace-Wells, D. 2019. *The Uninhabitable Earth: Life After Warming.* Penguin, Random House, NY.

- Weeden, D, & Palomba C., 2012. A Post-Cairo Paradigm: Both Numbers and Women Count. Pp 258-273. IN: Cafaro P. & Crist E., (eds). *Life on the Brink: Environmentalists Confront Overpopulation.* Univ Georgia Press, Athens, GA.

- Weisman, A. 2013. *Countdown: Our Last, Best, Hope for A Future on Earth?* Little, Brown, NY.

- Weisman, A. 2007. *The World Without Us.* St. Martin's Press, NY.

- Westen, D. 2007. *The Political Brain: The Role of Emotion in Deciding the Fate of The Nation.* Public Affairs Perseus Book Group), NY.

- Whitfield, J. 2006. *In the Beat of a Heart: Life, Energy and The Unity of Nature.* National Academy Press, Wash. DC.

- Whybrow, P. 2005. *American Mania: When More Is Not Enough.* WW Norton, NY.

- Wire, T. 2009. *Fewer Emitters, Fewer Emissions, Less Cost: Reducing Future Carbon Emissions by Investing in Family Planning; A Cost Benefit Analysis,* London School of Economics.

- Wilson, E.O. 2002. *The Future of Life.* Liveright Books, NY.

- Wilson, E.O. 2016. *Half-earth: Our Planet's Fight for Life.* Liveright Pub Corp, NY.

- Wilson. E.O 2019. *Genesis: The Deep Origins of Societies.* Liveright, NY.

- Wilson, E.O. (Ed.) 1988. *Biodiversity.* Nat. Acad. Sci. Press, Wash. D.C.

- Wilson, EO. 2014. *The Meaning of Human Existence.* Liveright Publ, NY.

- Wirth, T. 2009. Foreword: "A Pivotal Moment: Population, Justice and the Environmental Challenge." L. Mazur, (Ed), Island Press, Wash. DC.

- World Bank. 2005. *Environmental Fiscal Reform: What Should Be Done and How to Achieve It.* World Bank, Wash, DC.

- Wright, R. 1994. *The Moral Animal: Why We Are the Way We Are: The New Science of Evolutionary Psychology.* Vintage Books, NY.

- Wuerthner, G. 2012. Population, Fossil Fuels, and Agriculture., pp (123-129). IN: Cafaro P. & Crist E., (Eds), *Life on the Brink: Environmentalists Confront Overpopulation.* Univ Georgia Press, Athens, GA.

- Wuerthner, G. 2002. The Truth About Land Use in the United States. Watershed Messenger IX (2), online pub.

- Yergin, D. 2012. *The Quest: Energy, Security, And the Remaking of The Modern World.* Penguin Books, London.

- Yergin, D. 2020. *The New Map of Energy, Climate, and the Clash of Nations.* Penguin Press, NY.

- Young, R. & O. Pilkey. 2009. *The Rising Sea.* Island Press, Wash, DC.

- Zhang, H. et al. 2019. Microbial taxa and functional genes shift in degraded soil with bacterial wilt. Sci. Rep. 7:39911. doi:10.1038/srep39911.

- Zhou, Y. et al 2019. A practical, recyclable, ultra-strong, ultra-tough graphite structural material. Materials Today, 8 May, 2019. https://doi.org/10.10161/j.mattod.2019.03.018

Made in the USA
Columbia, SC
20 May 2024

9af1318c-0c6e-4a22-9612-86cf9a6676efR03